We Have Been Harmonized

We Have Been Harmonized

Life in China's Surveillance State

Kai Strittmatter

HARPER LARGE PRINT

An Imprint of HarperCollinsPublishers

HarperCollins books may be purchased for educational, business, or sales promotional use. For information, please e-mail the Special Markets Department at SPsales@harpercollins.com.

Originally published in Germany in 2018 by Piper Verlag GmbH.
First English publication in Great Britain in 2019 by Old Street Publishing.
English translation © 2019 by Ruth Martin

FIRST HARPER LARGE PRINT EDITION

ISBN: 978-0-06-302861-6

Library of Congress Cataloging-in-Publication Data is available upon request.

20 21 22 23 24   LSC   10 9 8 7 6 5 4 3 2 1

Contents

New China, New World   1
A Preface

The Word   21
How Autocrats Hijack Our Language

The Weapon   42
How Terror and Law Complement Each Other

The Pen   63
How Propaganda Works

The Net   81
How the Party Learned to Love the Internet

The Clean Sheet   145

Why the People Have to Forget

The Mandate from Heaven   176

How the Party Elected an Emperor

The Dream   195

How Karl Marx and Confucius are
Being Resurrected, Hand in Hand with the
Great Nation

The Eye   230

How the Party Is Updating Its Rule with
Artificial Intelligence

The New Man   297

How Big Data and a Social Credit System Are
Meant to Turn People into Good Subjects

The Subject   326

How Dictatorship Warps Minds

The Iron House   352

How a Few Defiant Citizens Are Refuting
the Lies

The Gamble   **364**

When Power Stands in Its Own Way

The Illusion   **379**

How Everyone Imagines Their Own China

The World   **389**

How China Exerts Its Influence

The Future   **447**

When All Roads Lead to Beijing

Acknowledgments   465

Notes   467

New China, New World

A Preface

The China we once knew no longer exists. The China that was with us for forty years—the China of "reform and opening up"—is making way for something new. It's time for us to start paying attention. Something is happening in China that the world has never seen before. A new country and a new regime are being born. And it's also time for us to take a look at ourselves. Are we ready? Because one thing is becoming increasingly clear: over the coming decades, the greatest challenge for our democracies and for Europe won't be Russia, it will be China. Within its borders, China is working to create the perfect surveillance state, and its engineers of the soul are again trying to craft the "new man" of whom Lenin, Stalin, and Mao once dreamed. And this China wants to shape the rest of the world in its own image.

The Chinese Communist Party (CCP) has placed its leader, Xi Jinping, where no one has been since Mao Zedong. Right at the top. Nothing above him but the heavens. China has a "helmsman" once more. Xi is the most powerful Chinese leader in decades, and he rules over a China that is stronger than it has been for centuries. An ambitious nation, readying itself to become even stronger—economically, politically, and militarily. The West's self-destruction has fallen into this nation's lap like a gift from the gods. With 21st-century information technology and its radical new possibilities for control and manipulation, the regime has instruments of power to which no previous autocracy has ever had access. Xi and his party are reinventing dictatorship for the information age, in deliberate competition with the systems of the West. And this has huge implications for the world's democracies.

Even within China, the CCP's plans are ambitious, but one shouldn't underestimate the hold that an autocrat has over his subjects' minds. The state has the ability to erase not just lives, but minds, in order to reformat them. The Tiananmen Square massacre in 1989, and the years that followed, provided a powerful demonstration of this fact. The date June 4, 2019, saw the 30th anniversary of the day the Chinese democracy movement was brutally crushed, and the Party has good reason to celebrate. In hindsight, its act of violence was a success—a greater success than anyone could have imagined at the time. The blood-letting gave the Party new life, as well as an opportunity to show what its mind-control apparatus could do, long before the advent of the digital age. Inside China, the memory of the massacre has practically been wiped out; the state-ordered amnesia is complete. And he who controls the past—the CCP understands this just as well as George Orwell did—also controls the future.

This is a message from the future, if things don't go so well. At the moment, things really aren't going well. That's why I wrote this book. It was born on the night Donald Trump was elected president of the United States of America, and was finished in the months that saw Xi Jinping "chosen by history," in the words of the journal of the Central Party School in Beijing, *Qiushi* (Seeking Truth). History is often a sluggish tide on which we float without ever being aware that it's moving. But that isn't the case right now: we are living through a time when the current of history seems almost physically tangible. Something is happening, to us and to China, and the two sides can no longer be separated.

The new age is one in which facts have been abolished; the Western world is suddenly mired in "fake news" and manipulated by "alternative facts." For me, though, there is nothing new about it. It's a life I've been living for twenty years, as a correspondent in Turkey (from 2005 to 2012), but above all in China. I studied in China in the 1980s, then worked there as a journalist from 1997 to 2005, and again from 2012 to 2018.

Government by lies is no doubt as old as the institution of government itself, yet we in the West are shocked by the return of autocrats and would-be autocrats to our midst, and with them the return of the shameless lie as an instrument of control. We had settled into the comfortable belief that these techniques and the political systems associated with them were obsolete. Autocrats everywhere are scenting an opportunity and joining hands with the populist agitators in our own countries. A perfect storm is brewing, for Europe and for democracies everywhere.

Xi Jinping has promised his people and the world a "new age"—and he is certainly building a new China. Both the Chinese people and the world at large have good reason to be nervous. Where Deng Xiaoping prescribed pragmatism, Xi Jinping has returned to revering ideology: he preaches Marx and practices Lenin with a force and dogmatism not seen for many years—and because he senses that Marx no longer speaks to many people, he has added Confucius and a fierce nationalism into the mix. Where Deng preached opening up and curiosity, Xi is sealing China off again.

Not that Xi is trying to force something on his party that goes against the grain. The opposite, in fact: he is fulfilling its most hidden desires with speed and precision. Until recently, more than a few Party cadres* were secretly asking themselves: what is it still good for, the Party—a vehicle for a long-dead ideology from a long-dead age, almost a hundred years old? But where the Party was starting to smell of decay, Xi gave it new strength and discipline; where it was stagnant and directionless, he breathed a new purpose into it. It thanked him by elevating him into the pantheon of its greatest thinkers during his own lifetime, and endowing him with almost unprecedented power.

Xi is now reminding everyone that this country was once conquered by the Party in a civil war. China itself was the Party's spoils of victory. In China, the army

---

* In its most general, military sense, a cadre is a unit of soldiers or officers. In the language of communism, however—and particularly the language of Chinese communism—a cadre is an individual Party official.

still belongs to the Party rather than the state. The state, too, belongs to the Party. And the Party—well, that seems to belong to *him*, now. It submits to the man who has given it a sense of purpose, and who is turning a one-party dictatorship back into a one-man dictatorship.

The Party calls Xi "the savior of socialism"—by which it really means "the savior of our power." The fate of the Soviet Union seems to trouble Xi deeply. He is quoted as saying that "what they lacked was a real man!" Not China, though. China has him now: Xi Jinping. For life. Today, hardly anyone is still prophesying the impending collapse of this system, and the Party can once again afford to think long-term. The year 2024 will be an epochal year for the Party. At that point, it will have overtaken the CPSU, its failed Soviet Union sister party, and the Chinese Communist Party will have become the longest-reigning Communist Party in history.

It is time for the West to let go of that form of wishful thinking that one wise author exposed as a "China fantasy"[1] some years ago: the idea that a more open economy and increasing prosperity would automatically bring political liberalization to China. For a long time, despite all the evidence to the contrary, people clung to the reassuringly pragmatic notion that if we engaged and traded with China, it would start to resemble us. After all, there was no precedent. This Communist Party was like none the world had ever seen. It simply assimilated capitalism and passed it off as "socialism with Chinese characteristics." It was an entity of phenomenal adaptability.[2] It never gave up its autocratic core, but in the past few decades, deep in the country's innards and even in the Party itself, there have been reform movements, original debates, surprising experiments, and brave taboo-breakers.

In Xi Jinping's China, this is no longer the case. He has brought unorthodox movements to a standstill. Xi the taskmaster is setting out to prove that an autocracy is better suited to making a country like China great and powerful; that the realization of his "China dream" requires a strong Party dictatorship. Xi is dispensing with the premises of Deng Xiaoping's policy of reform and opening-up; his China is no longer a state where everything is subordinate to economic success. Now, political control is at the heart of things. His Party is no longer one that devolves tasks to the state, to companies, to civil society, to the media, all of which have fought to carve out their own small freedoms. Xi has snuffed out those freedoms once again. During a single term in office, he has managed to get an iron grip on a nervous Communist Party stricken by a mood of crisis.

He took on a diverse, lively, sometimes insubordinate society and did everything in his power to "harmonize" it, as they say in China, stifling the voices of those who think differently and subordinating every last corner of society to the command of the Party. Xi, who claims to be incorruptible, is cleansing the country and the Party, including its ideology. He wants every last speck of land in China to be under his watchful gaze. Under Xi, the Party is becoming more godlike than it has ever been before.

Events on the edges of the Chinese "empire" have accelerated the new intransigence and addiction to control. In Hong Kong, hundreds of thousands of residents, fearing the permanent loss of their liberties, have taken to the streets. In Xinjiang, the party's abduction and indoctrination of probably more than a million Muslim Uighurs in a network of re-education camps is the largest internment of an ethnic-religious minority since the Nazi era. In China itself, the planned reprogramming of a people is evoking memories of the Cultural Revolution.

With one foot, then, Xi is taking a huge step backward into the past. Leninism is in his bones. And so is the thirst for power. Some compare him to Mao Zedong, but this comparison falls at the first hurdle: Mao was the eternal rebel, who thrived in chaos. In many respects Xi Jinping, who has a fetish for control and stability, is the antithesis of Mao. Xi is no revolutionary; he's a technocrat, albeit one who navigates the labyrinth of the Party apparatus with tremendous agility.

But one experiment from Mao's legacy is currently making a comeback: the CCP is once again practicing total mind-control, once again trying to produce "new men." Only this time, the Party believes that—at the second attempt—its chances are much better: China's dictatorship is updating itself with the tools of the 21st century. Because, with the other foot, Xi is taking a giant step into the future, to a place many dictatorships have sought, but none have yet found. The days when the Party eyed the internet with fear and anxiety are long gone. The regime has not only lost its fear; it has learned to love new technologies. China is staking more than any other country on information technology. The Party believes it can use big data and artificial intelligence (AI) to create steering mechanisms that will catapult its economy into the future and make its apparatus crisis-proof.

At the same time, it intends to use this technology to create the most perfect surveillance state the world has ever seen. Ideally, one where you can't even see the surveillance, because the state has planted it inside the heads of its subjects. This new China won't be a giant parade ground characterized by asceticism and discipline, as it was under Mao, but an outwardly colorful mix of George Orwell's *1984* and Aldous Huxley's *Brave New World*, where people devote themselves to commerce and pleasure and in so doing submit to surveillance of their own accord. Still, for the vast majority of subjects, the potential threat of state terror will remain ever-present, the background radiation in this Party universe.[3]

A central component of this new China, for example, will be the "Social Credit System," which from 2020 is intended to record every action and transaction by each Chinese citizen in real time and to respond to the sum of an individual's economic, social, and moral behavior with rewards and penalties. In this vision, omnipresent algorithms create economically productive, socially harmonized and politically compliant subjects, who will ultimately censor and sanction themselves at every turn. In the old days, the Party demanded fanatical belief; now, mute complicity will suffice. If the plans of Xi and the Party are successful, it will mean the return of totalitarianism dressed in digital garb. And for autocrats all over the world, that will provide a short-cut to the future: a new operating system that they can order in from China, probably even with a maintenance agreement.

Can this vision ultimately be realized in a country whose society is more diverse today than it has ever been, where the aspirations and consumer dreams of the new middle classes now hardly differ from those in other countries? Materially, at least, the Communist Party has delivered over the years. Under its rule in the past few decades, urban China has seen an unprecedented rise in prosperity. The Party has for a long time been making these middle classes into the country's most satisfied citizens, and therefore its greatest allies. Soon they may even be able to breathe easy: Xi Jinping has ordered the clean-up of the poisonous fog that still passes for air in China's cities. But the challenges are huge. Chinese society is aging rapidly, and Xi has not yet seen fit to try to bridge the country's divide between rich and poor. China, which calls itself communist, has long been one of the most unequal societies in the world. The number of billionaires in its capital, Beijing, overtook that of New York a few years ago, and its citizens are not blind to the fact that most of the extra money has been lining the pockets of a shameless kleptocracy[4] closely tied to the Party.

Xi's one-man rule comes with its own risks. A system that was until recently surprisingly adaptable is being made rigid once more, unreceptive to criticism and new ideas. His rule has created enemies and desires for revenge within his own ranks. Xi is aware of the problems. That is partly why he is giving his people the dream of China becoming the superpower that it was always supposed to be. He is also reintroducing an ideological enemy: the West. Of all the ways to unite the nation, nationalism is the cheapest. It's also the one that should cause the West most concern, because something else is also now a thing of the past: the idea of restraint in foreign policy. Xi Jinping has a message for the world: China is retaking its position at the head of the world's nations. And the Party media cheer: Make way, West! Make way, capitalism and democracy! Here comes *zhongguo fang'an*, the "Chinese solution."

After years on the defensive, under Xi Jinping the CCP is once more proudly proclaiming its system's superiority. China's democracy, says Xi, is "the most genuine democracy," and the most efficient to boot. The propaganda press crows that the liberal West is swamped by "crises and chaos."[5] "It is time for a change!" The self-destruction of the USA under Donald Trump is God's gift to the CCP in Beijing. So is a

Europe that has spent years absorbed in navel-gazing and family therapy sessions, no longer even noticing its shrinking significance on the world stage. We're not fighting a new Cold War yet, but all of sudden, competition between rival systems is back. Xi Jinping is now offering the world the "wisdom of China," by which he means the economic and political model over which he presides. What about this wisdom? Is China the country that has found a magic formula in the combination of autocracy and economic miracle? Advocates of this thesis point, for example, to the high-speed rail network that is already by far the world's largest. Or they contrast the giant airport in Beijing, which was built in record time, with the disaster of the seemingly never-to-be-finished airport in Berlin. They call it the China model, and they praise it as the model that will beat us hesitant democracies in the new competition of systems and in the end dominate the world. When, at the beginning of 2020, China built entire hospitals for the patients of the coronavirus epidemic in only two weeks, you could feel it springing up again, the admiration for the allegedly legendary efficiency of the Chinese system. The country had, after all, with a strong hand, managed to control the spread of the disease beyond the crisis region around the city of Wuhan and Hubei province. Quite a feat, especially when you contrast it with the disastrous crisis management of Donald Trump.

In fact, the coronavirus crisis exhibited the two narratives side by side as if in a magnifying glass: Is China the state unparalleled in mobilizing masses and resources for the common good, governed by a meritocracy drilled for efficiency? Or is it an internally brittle regime whose true nature was revealed once again here in the city of Wuhan: a state that, in the first decisive weeks, sacrificed without hesitation the welfare of its citizens to the Party's claim to power, with catastrophic consequences for both China and the world? Administered by bureaucrats in Hubei province who turned out to be as irresponsible as they were clueless. Dominated by a Communist Party which, today more than ever, is characterized by an excessive desire for control and secrecy. A regime which, in the face of crises, reflexively moves to cover up things, thereby recklessly encouraging the spread of the disease, particularly in the critical initial stages of the coronavirus outbreak.

In other words, a system that has robbed the nation of its social immune system—and which will ultimately always remain a risk to its people and to the world.

Mao Zedong, they say in China, vanquished the nation's enemies; Deng Xiaoping made the nation rich; and now Xi Jinping is making it strong, restoring it to its rightful position at the center of the world. With its "Made in China 2025" plan, the CCP wants to make China's economy a world leader in innovative technologies. And its "New Silk Road" project—the propaganda bureaucrats prefer the name Belt and Road Initiative (BRI)—is not just a global infrastructure and investment project, it's also part of the plan for a new international order more in line with the Party's ideas. China's goals are breathtakingly ambitious, it's true, but this country has taken our breath away several times before. China has long been the world's largest trading nation. In ten or fifteen years it will be the largest national economy on earth. In what other ways will China change the face of the earth?

And, crucially: how do we deal with it? In view of the lemming mentality with which many citizens of Western democracies have followed the pipes of right-wing populists and new would-be autocrats; and in view of the naivete and the blinkered attitude of many Europeans, who regard the comfort of their old world as God-given, I had an idea a while ago. People should be thrown out into the big, uncomfortable world whether they like it or not. It should be mandatory for all Europeans to spend a year living outside their comfort zones. They could be sent to Turkey, where democracy is being dismantled at lightning speed. Or to Russia, where lies and cynicism have long been the *modus operandi* of the state and of daily life within it. In my dream, people would then suddenly start to recognize things that are happening around them right now. And they would be brutally confronted with the logical endpoint of these things: tyranny.

Best of all would be to send them to China. In China, these Europeans would be lost for words at the ambition, the reckless pace of life, and the unshakable belief in the future, at the merciless competition of everyone with everyone else, and the untrammeled desire for wealth and power. The place would take their breath away, but perhaps also jolt them out of their lethargy and ignorance. It might give them the shock they need to stop allowing people in their own countries to divide them. In my fantasy, this experience provides them with courage, strength, and new ideas for the future in a humane, fair, and democratic Europe. As an added bonus they would eat incomparably better food in China than they do at home, and get to meet a whole range of wonderful, warm people, whose drive, energy, and courage is twice as impressive for the fact that it exists under a system like China's.

It's time for the democracies of the West to recognize

China as the challenge that it is. A confident, increasingly authoritarian China that is changing the rules of the game every day. This is not the China that the optimists once dreamed of: a country that might go down the same route as South Korea or Taiwan and, having reached a similar stage of economic development, set out along the path to democracy. It is a Leninist dictatorship with a powerful economy and a clear vision for the future: this China wishes to reshape the world order according to its own ideas, to be a model for others, to export its norms and values. And make no mistake: these norms and values are not "Chinese"—they are the norms and values of a Leninist dictatorship. China is creating global networks, increasing its influence. And the liberal democracies are being confronted with this new China just when the West is showing signs of weakness, and the world order it has constructed over the past few decades is sliding into crisis.

Of course the world can and should continue to cooperate and do business with China. But we need to do this in the knowledge of China's internal workings and its possible intentions. The Chinese model—the neo-authoritarian appropriation of the internet and new technologies—is not only working brilliantly, it's spreading: countries like Russia, Saudi Arabia, Vietnam, and Cambodia have long regarded Beijing as a role model, a trailblazer in the sophisticated manipulation of both the internet and its citizens. It was once said that capitalism would bring freedom to China. It didn't. Then it was said that the internet would subvert China's Party rule. At the moment, it looks very much as though China is subverting capitalism and the internet along with it.

We have good reasons to believe that democracy is better and more humane than China's system. But people often seem to forget one important thing: that although citizens of Western democracies may be living in the best of all times and the best of all places, such a life, free of violence and despotism and fear, is far from being the ordinary state of affairs in the long history of humankind. It was—and still is—a rather unlikely exception. Throughout human history, the overwhelming majority of people have lived in tribes, clans, kingdoms, and nations where chicanery and tyranny, corruption and despotism, persecution and state terror were part of everyday life. A vague sense that "it'll be okay" is no longer enough. In the past it has very often not been okay, and things are not okay on a lot of fronts right now. We in Europe and the United States should remind ourselves every morning: "It wasn't always like this. And it won't necessarily stay this way." Another reason to look to China.

This book is for those who, for whatever reason, are unable to spend their prescribed year in Beijing, Shanghai, Hangzhou, Chengdu, or Shenzhen. It is divided into three broad sections, though these sometimes overlap.

The first section explores the classic mechanisms of dictatorship: how it disconnects citizens from truth and reality, and how in the process it invents its own language. How it employs terror and repression when necessary, though propaganda and mind-control are its preferred methods, and why it must repeatedly inveigle its citizens into a collective amnesia. How it learned to love the internet: a first foretaste of the 21st century's possibilities.

The second section describes the reinvention of dictatorship in China. How the Party is creating a state the like of which has never been seen before, with the help of technologies designed to give the economy a turbo-boost and at the same time to dissect people's brains, exposing even their darkest corners. How China may soon overtake the USA in the areas of big data and artificial intelligence, and where it has already done so. Why the Party believes that, thanks to AI, it will soon "know in advance who is planning to do something bad"—as the Deputy Minister for Science and

Technology puts it—even if the person in question may not know it yet. Especially then. How the Party uses a "system of social trustworthiness" to divide people into trustworthy and untrustworthy, and plans to ensure that soon "all people will behave according to the norms." How it is already denying those who have betrayed its trust access to planes and high-speed trains. How, since time immemorial, dictatorship has produced warped minds rather than honest people.

Finally, the third section asks whether all this will work, and if so, what it means for us. It outlines the increasing influence that China's Communist Party has in the world, and how it is profiting from the weakness of Western democracies. And it explains why, in the end, the future will come down to whether we can rediscover our strength in time.

The Word

How Autocrats Hijack
Our Language

*"Enlightened Chinese democracy
puts the West in the shade."*

Xinhua News Agency, October 17, 2017

I live in a free, democratic country governed by the rule of law. I live in China. Yes, that's what it says on the banners and posters lining the streets in my city: Freedom! Democracy! The rule of law! I read this on every street corner in Beijing, every day. These are the "core socialist values" that the Party has been invoking for years.

Anyone who has lived under emerging dictatorships—in Turkey, Russia, or China, for instance—will be only too familiar with deliberate, systematic, and shameless perversion of facts. Donald Trump shows how you can apply that technique successfully in Western democracies if you are unscrupulous enough. His method is taken straight from the autocrat's handbook, in which lies are first and foremost an instrument of power. Fake News? Alternative Facts? To billions of people on this earth, they're an everyday, lifelong experience. I've spent two decades in China and Turkey: nations where left can suddenly mean right, up suddenly morphs into down. I was there as an outsider, an observer, always with the luxury of distance and astonishment at each new outrage. It's a luxury that a subject born into such countries can scarcely afford if he wants to get through life unmolested.

The Chinese have plenty of experience of rulers reinterpreting the world. Over 2,000 years ago, in 221 BC, Qin Shi Huang united the empire for the first time. His son ruled as emperor from 209 to 207 BC, with a feared and power-hungry imperial chancellor named Zhao Gao at his side. One day, in an audience with the emperor, the chancellor had a stag brought into the court. "Your majesty," he said, pointing to the beast: "A horse for you!"

The emperor was as taken aback as his ministers, and asked his chancellor to explain, if he pleased, how antlers could be growing out of a horse's skull. "If your majesty doesn't believe me," Zhao Gao replied, indicating the gathering of dignitaries around him, "then just ask your ministers." Some of the ministers were smart or scared enough to corroborate: "It really is a horse, your majesty." Of course, there were also those who stubbornly insisted that the animal standing in front of them was a stag. Later, the chancellor had them put in chains and executed. But he didn't stop there: whoever had remained silent in surprise or fear was also put to death. From then on, the stag was a horse. And a population had learned its lesson. *Zhi lu wei ma*—"to call a deer a horse"—is an expression in China to this day.

Western societies have grown comfortable in the certainties of the last few decades, and for the most part forgotten their experiences of the totalitarian systems of fascism and socialism. Thus the aspiring autocrat, equipped with an unscrupulous nature and a thirst for power, is always a step ahead of today's naive and un-schooled democrats.

When it comes to authoritarian personalities and systems, though, the primary intention is not to de-ceive, but to intimidate. That's why the lies of auto-crats are often shameless and outlandish. You may be a fan of Donald Trump or you may despise him, but there is no refuting what the whole world saw at his inauguration: a sparse gathering of onlookers on the National Mall, by reliable estimates about one-third of the crowd that had assembled for his predecessor's first inaugural address. Anyone watching, or reviewing the video and pictures taken of the event, would see that immediately. But the president, undeterred, has continued to vastly inflate the crowd size. Trump said there were a "million and a half people" in the audience. And his spokesperson even went so far to declare it "the largest audience ever to witness an inauguration, period, both in person and around the globe." In this respect, Washington is no different from Ankara. In a full-fledged autocracy, they would bus in those adoring hundreds of thousands; but in both cases the autocrat ultimately doesn't care whether people believe him. He doesn't want to convince everyone—but he does want to subjugate everyone. One essential feature of power is that, however great it becomes, it is never completely sure of itself. This paranoia, the fear of losing power, is part of the powerful man's nature. It's why he feels compelled to subdue the masses again and again. Above all, the lie serves this purpose.

If China's ruling party insists to this day that its country is communist, and if it is once again forcing teachers, professors, civil servants, and businessmen to make public commitments to Marxism, it isn't because it seriously thinks the population still believes in Marx. In the Swiss legend of William Tell, all the peasants were forced to salute a hat placed on a post by the imperial governor, Hermann Gessler. Marxism is China's version of Gessler's Hat: it is the gesture of submission that matters. This is how the autocrat deploys his lies—and refusing to swallow them marks you out as an enemy and a target.

But intimidation is only half the story. It's just as important to sow confusion, to disrupt the rationality and reality that give people a frame of reference, to take the compass away from the nation and the world. If you're a liar and a cheat, there's no way for you to win in a world that is repelled by these things, a world that differentiates between truth and lies. So you have to make everyone else a liar and a cheat, too. Then you will at least be *their* liar.

Hannah Arendt, who studied totalitarian regimes, said as much in an interview in 1974: "If everybody always lies to you, the consequence is not that you believe the lies, but rather that nobody believes anything any longer."[6] But a population that no longer believes anything is robbed of its ability to think and to judge, and ultimately of its capacity to act. As Arendt says, "with such a people, you can then do what you please." These are the ideal subjects—or the ideal opponents.

The mirror image of the liar's shamelessness is the

shame of the person being lied to, at least while he remains aware of the nonsense that he himself is bolstering every day, in chorus with everyone else. The act of repeating obvious untruths binds him to the liar with a rope of complicity. In the end, the ruler's lies breed cynicism among the people being ruled, who make their peace with the powerlessness of their situation and ultimately cling to just one thing: the leader's power. At that point, the leader no longer has to account for anything, because there is no truth left outside his fabrications.

In a world where the distinction between truth and lies has been abolished, there are just facts and alternative. The dominant values are not morality and a sense of responsibility, but usefulness and profit. If you do see the truth, it will do you no good to tell it; in fact, it's dangerous. Best of all is to acknowledge the lie as true and embrace it passionately—that's what the fanatics do. But they will only ever be a very small group. The next best thing is deliberately to avoid learning the truth, to live a life of benumbed ignorance—and if you do happen upon the truth, keep quiet and pretend you haven't. These two groups represent the majority of the population. Anyone who speaks the truth is either stupid or suicidal. The smart people in such a world are not the clear-sighted and wise; the smart people are the cunning and shrewd. There's no room here for common sense, or rather, ignorance is the new common sense, necessary for survival or used to justify opportunistic advancement.

Of course, the whole business of truth—recognizing it and communicating it through language—is philosophically difficult. "The name is only a guest in reality," said Zhuangzi, one of the forefathers of Taoism. Over 2,000 years later, the Nobel Prize laureate Herta Müller wrote: "The sound of the words knows that it has no choice but to beguile, because objects deceive with their materials, and feelings mislead with their gestures"; what counts when you write is "the honesty of the deceit."[7] Müller's "deceit" is well-intentioned; it participates in a free exchange with others' experiences, in full knowledge of the imprecision inherent in its claims.

Similarly, in a community, people make an effort to come closer to what is true, to gain a shared understanding of a world that looks slightly different to every individual. But the autocrat who claims sunshine when it's raining outside deliberately takes the world off its hinges. He creates a world according to his will, a world where things often mean the opposite of what they used to, a world in which balance can only be maintained if everyone huddles tightly around the leader.

And this leader often wants to create new men to go with his new world. From the outside, this world really does seem unhinged, in every sense of the word. Internally, though, it is structured in such a way that in the end, the last person to still believe that the earth turns around the sun will start wondering whether, after all, he's the madman. He will have to stop trusting his eyes, his ears, and his memory, and simply chew the cud of the information he's been force-fed.

For this reason, the free press is the autocrat's natural enemy. Where alternative facts are a badge of power, research and fact-checking by the free press equate to "ideological subversion" (as it says in the extraordinary "Document Number Nine," a battle plan by the CCP from 2013 to combat "Western values," to which we will return later).[8]

The autocrat who wants to create his own truth needs to conquer the word. In China there is no repression; there is simply "stability maintenance" (*weiwen*) and a "harmonious society" (*hexie shehui*). In the past decade, harmony has been one of the Party's favorite words: the harmony between orders and obedience. Harmony is when ordinary people don't make a fuss.

Take, for example, the "harmonious demolition" of houses by the city authorities to make way for property developers. In my little side-street in the center of Beijing, the city authority gave just a week's notice before it bricked up the windows and doors of all the snack bars, restaurants, hairdressers, newstands, and vegetable sellers, some of whom had been earning their living there for twenty years. The aim was to drive out the operators, since hardly any of them came from Beijing. This campaign was overseen by a dozen uniformed police officers, who protected the bricklayers from the displeasure of the street's inhabitants, beneath large banners that proclaimed: "We are improving the quality of life for citizens."

When China's president defended globalization at the 2017 World Economic Forum in Davos, he spoke of the increasing "opening-up" of China, while in fact his country was steadily sealing itself off. He invoked "global connectivity" while at that very moment China's censors were plugging the final gaps in the information blockade. What's more, he was applauded for it, because there is great confusion at the moment, all over the world. Some believe Xi, and some want to believe him. Some are blinded by his power. Some applaud because it is politically expedient, and serves their own interests to do so. China's power to twist words does not end at the country's borders.

It's a tried-and-tested tactic: steal your enemies' words and make them your own. As George Orwell taught us, freedom then becomes slavery, and ignorance becomes strength. And China is a democratic state under the rule of law. That's what the Party's propaganda says. And it's true: China does have a constitution, Article 35 of which guarantees citizens of the People's Republic "the freedom of speech, of the press [. . .] and of demonstration." There is a "parliament" in China, too: the National People's Congress. There are "elections," and citizens are regularly exhorted to make use of their "sacred and solemn right" to vote.

A long time ago, Lenin invented "democratic centralism": a system in which—so the theory went—democratically-elected functionaries should, once elected, have the privilege of dictating policy without opposition. Mao Zedong later preached the "democratic dictatorship of the people." In practice, centralism and dictatorship always ruled; democracy was a dead husk of a word that stuck in the throats of the population. The subjects of the regime thus experience their "elections," their "sacred right to vote," and their "freedom" as an eternal farce. The words lose all meaning; they have been discredited. In this way citizens are inoculated against subversive influences. When they come into contact with other worlds (a normal part of life for many Chinese people in our globalized age), they will not become infected by dangerous words that represent dangerous ideas. This perverted language makes the population immune. And mute.

The belief of the language-poisoners in the efficacy of their methods is by no means vain. Thought steers language, yes, but language can also steer and corrupt thought. "Words can be like tiny doses of arsenic," wrote Victor Klemperer, who explored the language of the Third Reich in his study *LTI* (*Lingua Tertii Imperii*): "They are swallowed unnoticed, appear to have no effect, and then after a little time the toxic reaction sets in after all." The language of dictatorship "changes the value of words and the frequency of their occurrence, it makes common property out of what was previously the preserve of an individual or a tiny group, it commandeers for the Party that which was previously common property," and in the process steeps words and groups of words and sentence structures in its poison. Making language the servant of its dreadful system, it procures it as its most powerful, most public, and most surreptitious means of advertising. In the end the Germans didn't need to consciously avow their belief in Nazism, because it had "permeated the flesh and blood of the people through single words, idioms, and sentence structures which were imposed on them in a million repetitions and taken on board mechanically and unconsciously."[9]

The autocrat's aim is to occupy and control the mind through language. The highest goal of the Communist Party's propaganda is to "unify thinking." But this is a process that has to be carried out over and over again. "We need to unify the thoughts and actions of all Beijing people," said Cai Qi, the Party Secretary for the capital, appealing to the propaganda press a few weeks before the CCP's 19th Party Congress in autumn 2017.[10] The totalitarian apparatus aims to unify all thought and action; every piece of "thought work" has this purpose. The goal is to flay the individuality from every individual, from his feelings, his judgment, his dreams. Only the "China Dream" is permitted now, with the Party as its artistic director. Individuals are supposed to merge in the great utopia, and their minds are being pressed into a new shape. Thus it was under Mao Zedong, and thus it is again in the China of Xi Jinping. It is no coincidence that one of the Chinese concepts to have made it into Western languages is "brainwashing"—xi nao in Chinese—invented by Mao Zedong's apparatchiks. To unlock brains, you need the right words. Stalin called writers "engineers of the soul." Like Confucius before him, Mao also knew that "one single (correct) formulation, and the whole nation will flourish. One single (incorrect) formulation and the whole nation will decline."

Of course, it isn't enough just to occupy the words of others. By the 1940s at the latest, China's Communist Party began creating its own new language for its new humans. Words that had fallen into disfavor were weeded out, and others invented to replace them. Immediately after the People's Republic was founded, Party linguists started work on the *Xinhua Zidian*, the *New China Dictionary*. Newly-minted politically- and morally-laden slogans and phrases have never stopped being fed into both the Party discourse and everyday language.

The language practice developed at that time still forms the foundation for what the sinologist Geremie Barmé calls "New China Newspeak"[11]—the jargons of various decades have been laid down in sedimentary layers, one on top of another. First the Marxist-Leninist imports were blended with the missionary, military swagger of the Maoist canon. Later, the wooden diction of Party bureaucracy was mixed with the technocrats' pseudo-scientific jargon. With Deng Xiaoping's politics of "reform and opening-up" and the increasing role of Chinese business in world trade, some bits of linguistic flotsam and jetsam from the worlds of commerce, advertising, and globalization floated into Party discourse—sometimes deliberately, sometimes by osmosis. And in the last few years, words plucked from the spheres of the internet and high-tech have been showing up to edify readers of leading articles in the *People's Daily.*

The emissions from the propaganda machine have become so saturated with this hermetic, opaque language that they have become indigestible to the people. Xi Jinping is not the first Party leader to combat "formalism" and "empty talk" in the ranks. In a famous speech in February 1942, Mao reprimanded his comrades for their "stereotyped Party writing"—they were filling "endless pages with empty verbiage." Such articles, Mao said, were like the "foot bindings" in which old women wrapped the broken bones of their tiny, bound lotus feet: they were "long as well as smelly."[12]

From time to time large sections of the population have enthusiastically adopted the Newspeak as their mother tongue, using it in private as well as public utterances. The Cultural Revolution of 1966–1976 was one such period. It was a cynical and brutal theater of power, set in motion by Mao, who was being marginalized by his rivals within the Party. He called on the country's young people to "storm the headquarters," to rebel against their teachers and professors, their parents, and the Party functionaries who cared more about orderly administration and a functioning economy than they did about permanent revolution. China's boys and girls burned for their messiah Mao, and they were ready to throw off all those trappings of civilization that were left to them: love of their fathers and mothers, and the last vestiges of feeling for their fellow man. By the time they had finished, China lay in ruins.

The very first thing they betrayed was the language of everyday life and common sense. "We were no longer humans," a former Red Guard soldier, now a lawyer, told me during an interview. "We were feral children, raised by wolves. A whole country, a whole generation that had suckled wolf's milk." His first name, Hongbing, means "Red Soldier." The most famous member of the Red Guard was the schoolgirl Song Binbin, a general's daughter who in August 1966, before the eyes of a million other young people, climbed the steps to the Gate of Heavenly Peace to be received by Mao himself. Song Binbin—thick glasses and braids—was permitted to fasten her red armband with its three characters, *hong wei bing*, Red Guard soldier, onto Mao. Mao asked the girl for her name; *Binbin*, as in polite and elegant? Yes, she said. "That is not good," said Mao. "*Yaowu* would be better: warlike." From then on, that was the 17-year-old's name. Poisoned milk, sucked in with poisoned words.

Some woke from the madness earlier than others. Young people like Gu Cheng, Mang Ke, Bei Dao, or Yang Lian were city-dwellers sent into the countryside by Mao. They knew nothing of each other, yet they were united by a common desire: to purify the language that had been beaten and gutted by propaganda, and fill it with new life. They did something unheard of, writing poems that used words like sun, earth, water, and death. The public, fed with nothing but slogans for ten years, was taken aback. Sun? Earth? Water? The young writers became renowned as the "Misty Poets" (*menglong pai*). In their poetry at least, the Chinese language was reborn in the People's Republic.

The gulf between official and non-official language is wider in authoritarian societies than in others. But because the private sphere is deprived of oxygen in totalitarian systems, people who live under them have official language forced on them at every turn. As a result, they develop split personalities—all the more so when the language of propaganda is the language of lies—and end up adopting what George Orwell perceptively called Doublethink and Doublespeak in *1984*: "To know and not to know, to be conscious of complete truthfulness while telling carefully constructed lies, to hold simultaneously two opinions which canceled out, knowing them to be contradictory and believing in both of them, to use logic against logic, to repudiate morality while laying claim to it, to believe that democracy was impossible and that the Party was the guardian of democracy."[13] Each subject acts a part—to his neighbors, his colleagues, the political apparatus—and as long as he is aware of this, he can still laugh or sigh about it in secret. For most people, though, the part they act quickly becomes flesh and blood, and because it is impossible to keep the two spheres perfectly separate, the language of the political apparatus always winds up corrupting the language of the people.

Writers have especially bemoaned the brutalization of the Chinese language by the militaristic, revolutionary battle-cry of Maoist times, the effects of which can still be felt today. Essays by the American literary scholar and sinologist Perry Link and the sociologist Anna Sun, for example, have looked at the legacy of Maoist jargon in the books of Mo Yan, China's first Nobel literature laureate. Anna Sun speaks of a "diseased language," one that Mo Yan, along with most of his contemporaries, has never outgrown.[14]

In everyday usage, the vocabulary of propaganda can travel a winding road. In the China of the late 1990s, *xiao zi* (or *petit bourgeois*), a group against which Mao had railed, suddenly became an aspirational term among the new middle classes: everyone wanted to be a *xiao zi.* In the new China, this was someone who could order a cappuccino in one of the recently-opened Starbucks; who knew that red wine should be drunk neat and not mixed with Sprite (as most of the Party functionaries and the nouveau riche did at their banquets); who sometimes took holidays to London and Paris. To be a *xiao zi* was suddenly cool.

From time to time, both words and citizens fight back. Many of the Party's "warlike" words have fallen prey to irony. *Tongzhi,* for example—*comrade.* Suddenly it wasn't just China's ardent communists who were addressing each other as *comrade*: it was also members of the gay community. Or the phrase *I have been harmonized.* For many years now, this has meant: *I have been caught by the censor, and my online comment—even my entire account—has been deleted.* When the police invite someone in *for a cup of tea,* it is interrogation rather than a hot beverage that awaits them. Sometimes a well-known intellectual, author, lawyer, or other inciter of unrest might *be traveled*: this creative verb-form denotes a person's involuntary removal from the city, while the Party has its conference or the foreign leader pays a visit.

China's propaganda incessantly spits out new words and phrases. Today's China is a fantastical realm of contradictions, a society rapidly branching out and exhibiting a pluralism that goes against the unification of all things and all actions so vehemently pursued by the Party. For such a country, the Party attempts to create terms that unite all contradictions, and thereby do away with them. "Socialism with Chinese characteristics" is one of these. Or the "socialist market economy." These formulations contain left and right, up and down, Maoist and neo-liberal all at once. Language has overruled logic and in doing so believes itself untouchable. Of course, in reality it is becoming ever more empty and absurd, but in a country where what matters is power and not letters, that doesn't really make a difference. Here, more often than not, the function of words is to convey an order rather than a meaning: Nod! Swallow! Forget! Kneel! And so the propaganda machine feels perfectly free to compare the Dalai Lama with Adolf Hitler, and at the same time to warn the country's newspaper editors never to confuse "truth and lies, good and evil, beauty and ugliness." The true, the good, and the beautiful are always the Party and its Word.

Naturally the Party doesn't stop at interpreting reality; it also creates it. "There are no dissidents in China." All you have to do is say it often enough. These words were spoken in 2010 by a foreign ministry spokesman in connection with the writer Liu Xiaobo, who had just been sentenced to 11 years in prison. In 2017, Liu became the first Nobel Peace Prize winner to die in prison since Carl von Ossietzky at the hands of the Nazis in 1938.

For more than a decade Liu Xiaobo was the most famous dissident in China. In official statements, though, he was always a "convicted criminal." What do you mean, *dissident*? "In China, you can judge for yourself whether such a group exists," the spokesman told journalists on February 11, 2010. "But I believe this term is questionable in China."[15]

At the time, the artist Ai Weiwei was still one of the most active Chinese micro-bloggers on Twitter. His analysis of this declaration appeared on his Twitter account:

1. Dissidents are criminals

2. Only criminals have dissenting views

3. The distinction between criminals and non-criminals is whether they have dissenting views

4. If you think China has dissidents, you are a criminal

5. The reason [China] has no dissidents is because they are [in fact already] criminals

6. Does anyone have a dissenting view regarding my statement?[16]

However, as Ai Weiwei was at that point a dissident himself, his blogs on China's own social networks had long since been deleted, and so hardly any of his fellow countrymen could reply. Twitter is blocked in China. Just one year later, Ai Weiwei spent three months in prison himself, supposedly for "economic crimes."

It seems fitting to conclude with a quote from Confucius, newly rehabilitated by Xi Jinping. Here's the philosopher's response to a pupil who asked what he would do first if he were handed political power: "He who would create order in the state," replied Confucius, "must do one thing: correct the names."[17]

The Weapon

How Terror and Law
Complement Each Other

"Political power grows out of the barrel of a gun."

Mao Zedong[18]

This is one of Mao's most-quoted pronouncements. But what people often forget is that, alongside and equal to the barrel of the gun, Mao and his people always had the barrel of the pen—propaganda. The Maoists used to mention the two in the same breath: "The Revolution relies on guns and pens." One stands for the threat of physical violence and terror; the other for mind control. As *wu* (the military; force of arms) and *wen* (literature; culture), these two instruments have served the rulers of China as far back as the classical period.

Once victory had been achieved in the civil war, the pen quickly became the weapon of choice. Today's Party still commands both: the pens and the guns—and in China, as I have mentioned, even the People's Liberation Army answers to the Party, not to the state. The Party has the monopoly on authority, and it exercises this authority over the life of every individual. The people's awareness of this fact is refreshed on a regular basis, not least with broadcast images of massed troops. For months after the event, on the screens of every metro car in Beijing, the video of the great military parade that took place in the heart of the capital in September 2015 played on a continuous loop from early morning to late at night.

The actual use of force against political undesirables is limited to exceptional personalities such as human rights lawyers and prominent dissidents, and to exceptional situations, such as crushing severe unrest in the provinces (most often Xinjiang or Tibet) or the Tiananmen Square massacre in 1989. Millions of students, workers, and citizens from all walks of life had spent months demonstrating against corruption and abuse of power, before the tanks rolled through Beijing on June 4.

Mao Zedong had no equal when it came to terror, whether inflicting it on the people or the Party. The climax of this policy came with the Cultural Revolution, when Chinese people had to fear not only the ruler's henchmen but each other—their own husbands, wives, and children. In the decades of reform and opening-up that followed under Deng Xiaoping, with the gradual emancipation of society, fear of the omnipotent, despotic Party receded. People could breathe again, and they enjoyed new freedoms. Yet the Party made sure the memory of its power never completely disappeared. When it discovered to its horror that emancipation was generating independent thought and actions, that the seed of something like pluralism was germinating in China, with nature conservation societies and religious charities, feminist circles and legal aid groups springing up, the Party took fright—and once again brought out the instruments of repression and enforcement.

Even in the years before Xi Jinping took office, one could observe how the apparatus of repression was being reinforced. In 2012—the year I returned to China—the government presented the third budget in a row in which spending on internal security outstripped the national defense budget. That included funds for the police and the courts, as well as the state security apparatus, which had expanded hugely under Zhou Yongkang. Under Xi Jinping, spending rose again: the German sinologist Adrian Zenz calculated that in 2017, the budget for internal security outstripped military spending by almost 19 percent.[19] The state and the Party are now arming themselves more against domestic than foreign enemies—and with an astounding attention to detail, as became evident before the festivities for the 65th anniversary of the founding of the People's Republic. "10,000 pigeons go through anal security check for suspicious objects," reported the *People's Daily* on its English-language Twitter page, beneath a photo of an innocent-looking white pigeon.[20]

Under Xi Jinping, the Party's re-inflamed hysteria over security has assumed entirely new proportions. Immediately after taking office, Xi got to work plowing back into the ground all those colorful shoots of civil society that had sprouted over the preceding years. He silenced the internet once more, and the press, which had begun to develop opinions of its own. Large numbers of bloggers, authors, and intellectuals fell silent; some disappeared altogether. Xi had draconian new laws drawn up, while purging with ruthless efficiency those who over the preceding years had tried to use China's legal code to protect its citizens from the state. Civil rights lawyers became the focus of a campaign in which they were terrorized, slandered, locked up, broken, and then paraded before the public.

The rule of fear has returned under Xi Jinping. Before 2012, the regime tried to keep its repression hidden, but when Xi took the reins the weapons went proudly back on display. The new leader also sowed fear within the Party: his anti-corruption campaign has been impressively long and harsh, but in the end it is still just that—a campaign. It hasn't ushered in constitutional reforms, and it doesn't dare grasp the root of the evil. There is no prospect of independent oversight—by the media or the judiciary, say—of a Party apparatus that continues to treat the country as a giant self-service store where nothing has to be paid for.

The source of Xi's power over his own ranks remains intimidation. And that is certainly effective in the short term: during his first term in office, Party functionaries all over the country were paralyzed with fear of the dreaded Central Commission for Discipline Inspection—one of the country's most secretive and powerful organizations—and the suicide rate among CCP workers doubled. Between 2009 and 2016, according to a study by the Institute of Psychology at the Chinese Academy of Social Sciences, 243 Party officials took their own lives (140 jumped to their deaths, 44 hanged themselves, 26 took poison, 12 drowned themselves, and 6 slit their wrists).[21] These figures are likely to fall short of the true number.

The Discipline Inspection Commission doesn't just

fight corruption; it also investigates comrades' ideological loyalty. Its inspectors have the power to end careers and to spirit people away into secret interrogation rooms for months. Like a modern-day Spanish Inquisition, it has hunted down dissent in the country's ministries and state companies, in the universities and think tanks. Formerly noisy advocates of passionate and controversial arguments have suddenly started to keep a low profile. Many who can afford it are going abroad or at least trying to send their money overseas, so that their families—and especially their children—have a way out.

The Commission also goes after dissidents in society at large. Civil rights lawyers, for example, must live with the constant surveillance and intimidation of family members, friends, and landlords. They are regularly summoned for interrogation, placed under house arrest, put into secret or official prisons. Some vanish without a trace for long periods, are tortured, or locked away in psychiatric hospitals.

Beijing's paranoia that the "Jasmine Revolution" in the Arab world might spill eastward sparked the first wave of repression against human rights lawyers in 2011. Even at that time, arrested lawyers were made to "confess" and "repent," sometimes using torture. Eva Pils, a legal scholar and China expert, quoted one victim as saying: "Not only did they want to make you say *that* black was white, you also had to explain *why* black was white."[22]

The Party feels the need to strike terror into anyone who challenges the state—even when that challenge simply means taking it at its own word, by pointing to the country's laws and its constitution. At the same time it likes to keep the level of fear among the general population topped up, so that everyone understands that the single, weak act of dissent permitted in China is the sigh, the shrug, and the murmured words *mei ban fa*—what can you do?

This is the reason behind all the public spectacles, and it can be summed up by the Chinese expression "killing the chicken to scare the monkeys." For decades, these have included the show trials of political opponents and dissidents that have become a regular feature on the national evening news. Now President Xi Jinping has given his security and propaganda apparatus a new weapon: the televised confession. The public shaming sessions, which were a central part of the Cultural Revolution's reign of terror, have been reinvented for the media age.

Since the summer of 2013, the state broadcaster CCTV (China Central Television) has been delivering to its viewers a never-ending parade of people who have been arrested or previously "disappeared." In the course of "interviews" mostly filmed in prison, they play the role of the repentant sinner confessing to their misdeeds—long before they ever get to see a lawyer, let alone the inside of a courtroom. These people might include a weed-smoking pop star, a murderous cult leader, or an economics journalist like Wang Xiaolu. In the wake of the stock-market crash of 2015—and the government's embarrassment over its rash, panic-stricken reaction—Wang was set up as a scapegoat. On television, the penitent journalist confessed that by writing a "sensationalized" article, he had been single-handedly responsible for the panic on the financial markets.

There have been appearances by a popular liberal blogger, a critical publisher, and an employee of a Swedish NGO who advised human rights lawyers. They all display repentance and humility before hundreds of millions of viewers, make tearful confessions about their "criminal" activities, denounce their former colleagues, expose their backers in "hostile foreign countries," thank the Party for its benevolence, and beg for a second chance. The scripts are often the same, right down to individual phrases.

In a book published in 2018, twelve former prisoners tell the story of how these interviews happened.[23]

One was first forced to take medication, before being filmed through the bars of his cell reciting sentences he had learned by heart. Another recounted how officials would instruct him when it was time for him to weep. Wang Yu, one of the country's best-known civil rights lawyers, reports how the officials at her interrogation showed her a photo of her son, who was 16 at the time. He had been labeled "suspect." Wang fainted, and when she came round, an interrogation officer was standing beside her:

> He told me that my son had been captured by anti-China forces but that luckily the police had found him and he was currently in Yunnan. He said my attitude would decide whether my son would be saved. I didn't know what to feel. I asked: how could I save him? He said that I should record a video for the PSB* boss to demonstrate my [good] attitude. I asked: What kind of video? What kind of attitude? They wrote down everything that I had to say on a piece of paper and asked me to memorize it. I don't remember clearly what it said, just that it was about denouncing certain anti-China forces.[24]

---

* Public Service Broadcasting: in other words, Chinese state television.

At no point did Wang Yu have any idea that her "confession" would later be aired on television.

For some in China, the practice awakens memories of dark times. "This is an echo of the Cultural Revolution," Beijing lawyer Si Weijiang told me. The breaking and public humiliation of dissidents, the self-criticism and regret displayed before the masses—all this was a daily reality for the Chinese people during the terrible decade of the Cultural Revolution. At that time, the "struggle sessions" would take place on public streets and squares; a teacher once tormented in this way compared the experience to a gang rape. The CCTV confession is the resurrection of this practice; less bloody, it's true—during the Cultural Revolution, some did not survive their run-ins with the mobs summoned to humiliate them—but effective nevertheless.

Civil rights lawyers have been among the groups hardest hit by this media humiliation: the unprecedented arrest of lawyers in the summer of 2015 was accompanied by perhaps the largest smear campaign in recent history. In the eyes of the Party, the lawyers were a growing problem. Their number had increased nationwide to two or three hundred; they were connected; they were pooling their resources and know-how; and they were savvy in the way they used their online audience to press their cases.

Following the arrest of almost all the lawyers from the busy Beijing Fengrui law firm—really a "criminal organization," so the presenter revealed—CCTV offered viewers an extraordinary spectacle. The lawyers paraded before the cameras not only incriminated themselves; they also vied with each other to denounce the others. Assorted video clips were interspersed to spice the whole thing up with a hefty dose of sex and character assassination. The lawyers, who had often acted for ordinary people—mostly the victims of state officials' despotism—were portrayed as morally bankrupt, greedy hustlers. The head of the practice, Zhou Shifeng, CCTV revealed, had enjoyed six lovers at the same time. It even created an infographic that displayed them all from one to six.

The message is clear. The Party can transform any lawyer into a criminal overnight, any hero into a nobody. And the underlying motives are also clear: as always, the goal is both to intimidate the target group, and at the same time discredit them in the eyes of the rest of society. Does it work? "I'm afraid so," says the lawyer Si Weijiang. "I'm sure most viewers believe what they see."

The tactic works even on those who don't fall for the spectacle, and see it for the theater of the absurd that it is. The staged confessions are so hair-raising, and make such a mockery of every last semblance of the rule of law, that they serve as a persuasive demonstration of the all-powerful, despotic state. Only a lunatic would dare go up against it.

Lawyers like Si Weijiang who are still free, and have not yet had their license to practice revoked, call the CCTV confessions a "travesty of the constitutional state." Yet this description assumes that there is a constitutional state to be travestied—that China should be measured according to the same yardstick as countries with a functioning justice system and a genuine separation of powers.

For a few years, it's true, there were great hopes, including among China's jurists themselves, that the country's justice system might move in this direction. Hope spread to the West, too, with phenomena such as the German-Chinese "rule of law dialogues," which have taken place annually since the year 2000 (providing, among other things, a convenient screen behind which a Western political establishment primarily interested in business opportunities can hide). Today this hope is buried deep, though seldom has a Party leader mentioned the rule of law so often as the present one. Xi has publicly scolded Party functionaries who don't give proper attention to the law. "We will spread the rule of law throughout the country," he once announced. For the CCP, though, the rule of law means something completely different from what it means to most citizens of Western democracies.

The Chinese word for rule of law is *fa zhi*, and is made up of the words *fa* (law) and *zhi* (rule). China-watchers spent many years puzzling over where the Party might take this idea. Would they tread a slow path toward the "rule *of* law" in our sense? Or was the destination to be "rule *by* law"—with laws mere tools in the service of power? The riddle has long since been solved. Xi Jinping himself has compared the role of laws to the "handle of a knife in the hand of the Party." In summer 2015 the lawyer Zhou Shifeng, head of the above-mentioned Fengrui practice, explained to me his interpretation of the Party's "rule of law": "What they mean by that is: 'I will take my laws and rule you with them.'" Less than four weeks after our conversation, Zhou was in prison. Shortly afterward he made a forced confession on state television, where he was shamed as the mastermind of a "criminal gang"; and a year later he was sentenced to seven years in jail for "subversion." To be subversive in China, it is often enough that you have tried to take the Party and its laws at their word.

China's legal authorities have traveled a long and remarkable route to reach this point. After the murderous chaos of the Cultural Revolution, the CCP under Deng Xiaoping made a public commitment to build a functioning justice system. In 1982, the country drew up a new constitution for itself. The CCP always made it clear that it would not tolerate any direct challenge to its power, but little by little Chinese society became freer and more confident. Economic development brought with it not just prosperity but myriad new conflicts. More and more citizens were becoming aware of their rights, and more and more lawyers were prepared to defend them, including against the despotism of officials. They made use of their freedoms, and sometimes they stretched them. Barely a decade ago, the media in China—including the state media—was still celebrating people who are today in prison, hailing them as heroes in the fight against injustice. Of the fourteen Chinese lawyers whom the Hong Kong–based magazine *Asia Weekly* crowned "people of the year" in 2005, not one has since managed to escape arrest or mistreatment by the authorities.

Meanwhile, China has passed a slew of wonderful new laws in a range of areas. It's just that no one pays any attention to them. "Being denied access to the court files, and to my client, is an everyday occurrence," says one lawyer. A report by a Beijing insider says that justice staff are often "rude, arrogant, overbearing, and disinterested." During interrogations, "rules are broken" and "force" is often used. The judges "don't abide by the law; they don't even listen to the victim and the lawyers." They also "take bribes and pass their judgements on the basis of relationships." The author of this damning account is no outspoken critic of the regime, but a senior civil servant within the highest state prosecutor's office in Beijing.[25]

In criminal trials, China has a conviction rate of over 99.9 percent: if you've been arrested and charged, you've pretty much already been found guilty. Acquittals are as rare "as the feather of a phoenix and the horn of a unicorn," because the police and the state prosecutors are infallible. In this system, lawyers standing up for the civil rights of their clients have always had a hard time. A report published in Hong Kong documented dozens of cases between 2006 and 2015 in which imprisoned lawyers had been tortured.[26] Among other things, the victims reported suffering electric shocks, burns, maltreatment of sexual organs, and sleep deprivation. And yet their numbers continued to grow—until Xi Jinping's wave of repression rolled over them.

Once upon a time, say those who are still at liberty and willing to talk, they put China's lawyers behind bars. Then it was the lawyers' lawyers. Now they're locking up the lawyers' lawyers' lawyers. In the end, what has happened under Xi Jinping is more than just an attempt to keep these lawyers in check: it's an attempt to discredit the entire movement for legal reform, and ultimately to wipe it out.

Xi Jinping himself has sent out seemingly contradictory signals on the matter. On the one hand, in his speeches he acknowledges that a modern economy and society require an effective justice system. On a local level, he wants to suppress the influence of Party functionaries on courts. In addition—following a years-long campaign by those very lawyers he is now persecuting—he has done away with the *lao jiao* re-education camps, into which people could vanish for up to four years without ever seeing a judge or a lawyer.

At the same time, the Party chief is a hostage of the system he has sworn to defend. Instead of re-education camps, unpopular citizens are now vanishing into other facilities. A genuinely independent justice system is a thing of horror for Xi. And so, on the same day the Xinhua News Agency praises the "rule of law," it runs a story saying that even to discuss "whether the Party or the constitution is the higher authority" is "absolutely forbidden."[27] One might as well call it blasphemy, in fact, given that the Party stands godlike above everything.

Ah yes, the constitution. Article 35 states: "The

citizens of the People's Republic of China enjoy the freedom of speech, of the press, of assembly, of association, of procession, and of demonstration." In reality, though, Article 1 is the only one that matters: "The People's Republic of China is a socialist state under the people's democratic dictatorship." And: "Sabotage of the socialist system by any organization or individual is prohibited." The very first group of citizens to fall victim to Xi Jinping's autocratic instincts, at the very start of his first term in office, was the movement headed by the rights activist Xu Zhiyong. It had demanded something unprecedented from the state: namely that it abide by its own constitution, including all the fine-sounding articles that came after the first. That demand led to show-trials, house arrests, and custodial sentences of several years.

The Party's schizophrenic tendencies were on full display when it declared December 4, 2014, to be "Constitution Day." Yet Weiboscope, a project set up by Hong Kong University to document censorship, found that the most blocked and deleted word on China's social networks on the new Constitution Day 2014 was "constitution." It was fundamentally "wrong to say 'rule of law' contradicts the Party rule," wrote Wang Zhenmin, dean of the law faculty at Beijing's Tsinghua University, in the *People's Daily*. "Law in China," he explained with striking honesty, was "nothing but the codification of the directives of the Party."[28]

The debate around the need for an effective judiciary during the reform years focused on China's judges as well as the lawyers. Here, professionalization was the buzzword: judges should be better paid and properly trained. In judicial circles, Western standards began to be discussed, often approvingly. The ideal of independent courts was mentioned with growing frequency in official documents. Some judges openly declared that political interference in important judgments had to stop.

For many years, China's highest judge, Zhou Qiang, was among the campaigners. By the start of 2017, though, even this long-standing reform optimist could see the writing on the wall, suddenly launching an attack on the independence of the judiciary and warning that China must not fall "into the trap of a false Western ideology." China's judges and legal minds must "draw their swords" against harmful influences such as "the separation of powers" or "the independence of the judiciary."[29] On Weibo, the largest Chinese microblogging service (Twitter has been blocked since 2009), China's supreme court added helpfully: of course there is judicial independence in China—but "only under the leadership of the CCP." Which is not unlike telling a prisoner to enjoy his freedom, just as long as he stays within the prison walls.

Under Xi Jinping there has been a flood of significant new laws, whose main purpose seems to be to give a *post-facto* legal underpinning to earlier acts of despotism. "Once I was walking out of the office with a colleague," the Beijing lawyer and civil rights activist Li Xiongbing told me, "and he was dragged into a car by police officers, interrogated and beaten for seven, eight hours and released again the next day. And they didn't give him a single reason why. That doesn't happen anymore. Of course, they still take people away just like they used to—but now they have to think up an excuse, name an article of the law." You might call that progress. But you might also be reminded of Joseph Stalin's assertion that "We need the stability of the laws now more than ever." That was in 1936, in the midst of the purges that have gone down in history as the "Great Terror."

It is nothing new for China's autocrats to view the law mainly as a method of control: a way to subjugate the population and to keep them down. Nearly 2,500 years ago, philosophers like Shang Yang and Han Feizi founded a school of government on the idea that strict control of the people through laws was the necessary basis for a united empire and central state. All of China's subsequent emperors have drawn on these ideas, outwardly acting in accordance with Confucianism and paying lip service to morality and virtuous governance, while in reality the machinery of state was driven by a draconian catalogue of laws, decrees, and threats of punishment. When they need to justify authoritarian rule, the propaganda merchants still like to quote Sun Yat-sen, father of the 1911 Revolution. The Chinese, he wrote, are "like a heap of loose sand": billions of little grains with nothing holding them together.

But this reasoning is backward. Chinese society does not eternally require a strong state because of some inherent inability to govern itself. It's the reverse. The Chinese state deliberately keeps society crippled. "When the people are weak, the state is strong; when the people are strong, the state is weak," goes a saying from *The Book of Lord Shang*. "Hence the state that possesses the Way devotes itself to weakening the people."[30] Before he can turn to his enemies, writes Shang Yang, it is a ruler's first task "to overcome his own people." He goes on: "The root of overcoming the people is controlling the people, as the metalworker controls metal and the potter clay. When the roots are not firm, the people will be like flying birds and running animals. Who will then be able to regulate them? The root of the people is law. Hence those who excel at orderly rule block the people with law."[31]

This concept of law and order is still useful to the autocratic rulers of today. But even in China's hybrid system, where an elaborate apparatus of laws provides for the efficient administration of a modern state, despotism remains at the core. The corridors of power may be wallpapered with legal documents and articles, but they can be torn down at any time. The autocrat has no interest in the genuine, consistent rule of law; his ideal is to keep his subjects in suspense at all times, always uncertain and afraid.

Laws in regimes like China's are formulated in terms so vague and contradictory that practically every citizen breaks one of them at least once a day. For the overwhelming majority of people and in the overwhelming majority of cases, there are no consequences to this habitual—and often necessary—law-breaking. Under a system where such hypocrisy is rife, it isn't surprising that most citizens take a somewhat jaded view of the law. But if the state has set its sights on you, then whenever it wants to strike, it has the appropriate article and clause ready for use at any time. These clauses are the hooks upon which it hangs disobedient citizens for all to see.

The Pen

How Propaganda Works

"History has proved and will continue to
prove that the Communist Party of China
is great, glorious, and correct."
People's Daily, *April 11, 2016*

In everyday life, propaganda is the principal instrument of government for the Party. It would be a mistake to believe that this propaganda, which to outsiders often comes across as crude, vacuous, and absurd, doesn't work. Much propaganda *is* crude and vacuous—and it works surprisingly well. Most important, it is omnipresent.

Just a few months after Xi Jinping took power, new posters appeared everywhere. On the large crossroads near my house, propaganda posters suddenly lined the streets for hundreds of meters. The design of the posters was unusually fresh and colorful; many were based on the aesthetics of traditional folk art and peasant painting.

One of these posters showed three young, cheerful peasant girls decked out with flowers. Above the image, in bold characters, were the words "Sing the Party a folk song." It's the title of a song that became popular during the Cultural Revolution, but was taught to generations of children afterward, too. "The Party is a mother to me," read the other lines on the poster—another verse from the song. It continues: "My mother only gave me my body, but the Party lights up my heart." Notice the word "only": it tells you that the Party deserves greater thanks than your own mother. The Party protecting the people and feeding them at its maternal breast is a trope that has been kept alive to this day. "Know the goodness of the Party, feel the goodness of the Party, follow the Party and ensure stability," says a banner a few meters farther down the street.

The Communist Party is not just Big Brother. It's Big Mother too, according to German expert on China Bernhard Bartsch. Thousands of speeches over the decades have been made on the "warm-and-full question": that is, the Communist Party's stated aim that every Chinese citizen should be able to eat his fill and dress himself in warm clothes. This, the Party proclaimed, was the highest human right, and all other human rights would have to wait in line until that had been achieved. One of the most-quoted passages from the great Party conference speech by Xi Jinping is the sentence: "No one must be left behind." Every last person in China should feel the generosity of the Party.

If the Party is given the role of mother, then the Party leader fulfills the role of father. In this household, the regime's subjects are children who enjoy a claim to protection and affection, but also require instruction and a firm hand. The Chinese language itself reinforces this narrative: the word for the state, *guojia*, literally means the "state family."

At the start of Xi Jinping's second term in 2017, state television put out a short video entitled "Family, State, World." The word they used for "world" was "everything under the heavens," *tianxia*, the term once used to describe the geopolitical sphere controlled by the son of heaven, the emperor. Xi appears in this video as a benevolent patriarch, who lets those under his protection share in his charisma and his care, by passing on "the warmth of love with his two hands." The whole empire is a family, and Xi is the just, self-sacrificing head of the household. At one point in the film, he is sitting with elderly peasants in their hut, and as he takes the hand of one peasant woman, a blazing beam of light shoots from her heart to his hand. "The family is the smallest state," CCTV captions the image. "And the state is made up of millions of families."

This is a direct reference to the Confucian classic *Daxue* (*The Great Learning*), written more than 2,000 years ago. It not only places a duty on the ruler to govern virtuously and remain in touch with the people—it also places a duty on subjects to play their part in helping the world function. Ultimately, every individual is responsible for the order of the universe. Only he who cultivates himself and who lives honestly and virtuously will bring his personal life into order; and only then can he regulate his family. And only when families are regulated can the country be governed well. As long as everyone plays his part, peace will reign on earth— *tianxia taiping.*

It's an old idea among Confucians: that shaping people through education serves state order, which in turn serves heavenly order. People can learn to be good, to think, speak, and act correctly. They learn from role models, ideally from a virtuous ruler; they can perfect their personalities through tireless work on themselves; but they can also be shaped by "rites." (The work canonized by the neo-Confucians of the

Song dynasty as *The Great Learning* was originally a part of the *Book of Rites*.) The Soviets, too, wanted to create "new men," but Mao Zedong drove the project forward in China with an even greater passion. He set about putting his subjects—and not just the ones in the re-education camps—through what he called "thought reform." And as he prepared to give socialism the gift of his new men, free of all ties and ready to serve the revolution selflessly, he was able to draw on the ideas of the Confucians he so hated.

At first glance the brightly-colored posters that transformed public spaces over the whole of China shortly after Xi Jinping took power looked very different to the ever-present red slogan banners of the previous decade. Some praised Party Leader Xi Jinping's "China dream," while others warned people in good Confucian style to respect their aged parents. The majority, though, drew the reader back to the People's Republic of yesterday via the familiar, stale slogans that were printed across the cheerful images. In this land, "the people are happy" (thanks to the Party) and "China is strong" (thanks to the Party), and "harmony" reigns, even among geese and chickens (thanks to the Party).

Unlike other propaganda campaigns, however, this one had a long tail; the posters are still there years later, even as I write. They are regularly replaced and occupy new building-site fences and neighborhood walls every day. Their images have now made it onto Air China's on-board screens. Before take-off, passengers are reminded to obey their elders and to recall other Chinese virtues. The propaganda posters have multiplied, while commercial advertising, which was still dominant until recently, has noticeably retreated from the streets of Chinese cities. Power is now showing its true colors; urban spaces have turned red again.

After their initial surprise, most people simply ignored the flood of posters, but some were quite baffled by them. The writer Murong Xuecun, for example, tells the story of how he met an old friend, a Party functionary, and asked him if his Party knew what year they were living in now. Did they really think they could still get through to people with such hackneyed slogans? Of course the posters were stupid, his school-friend acknowledged. "But that doesn't matter." Then he explained the real message: "We can cover the walls with this stuff. Can you?"

The implication was: "This is how great our power is. The whole world around you belongs to us. We are going to wallpaper your heaven and your earth. And you are only a guest here by our grace." This isn't only about the words, it's about overpowering people.

Haifeng Huang, a political scientist at the University of California, calls it "hard propaganda." He carried out a field study in China which came to the conclusion that such propaganda could "worsen citizens' opinion of their regime" while at the same time fulfilling its purpose: "signalling the state's power and reducing citizens' willingness to protest."[32]

The lives of China's 1.4 billion citizens are saturated in the Party's teachings. The indoctrination starts with the counting rhymes and patriotic songs they hear at preschool; it is the corset into which the school curriculum is laced, and it forms the basis of the political training that staff in the civil service, universities, and state-run companies all have to attend on Friday afternoons, to explore the latest twists and turns in "Xi Jinping Thought." In rural villages, it appears in large characters painted on walls; city-dwellers encounter it on the noticeboards of their street or neighborhood, on the red banners that hang from apartment blocks and road bridges, and in some cities even on scrolling LED screens mounted upon the roofs of taxis, which repeat the Party's latest slogans again and again: "Do not forget our original mission!" The message of the CCP reaches you in the country's museums and on the old sites of the Revolution, which have been restored over the last few years to serve "Red Tourism," and to which schoolchildren, workers, and company employees are now bussed in their millions. It permeates the radio, newspapers, internet, television, and cinema.

In a post on the messaging service WeChat (*Weixin*)—swiftly deleted—the sociologist Sun Liping from Beijing's Tsinghua University identified three techniques for "mind control."[33] One central technique is the control of news sources: "The meal you cook can never be better than the rice you cook it with." The system successfully blocks information from outside and replaces it with "patriotic education." Hence, for example, the ubiquitous narrative in which China's "special national circumstances" have made the country into a unique place unlike anywhere else in the world, and which requires the Party to rule in the precise way China's subjects are currently experiencing.

Second, the system starts building the parameters for your thought when you're very young, changing the way in which you ask questions and steering you into predetermined channels. Once you have swallowed and internalized what the Party has fed you, says Sun Liping, you can't even ask certain questions: they lie outside your realm of experience and powers of imagination. And third, the system inspires the kind of fear that suppresses awkward questions: "If you don't swallow all this, you'll be punished."

The incessant bombardment of people with these messages is a crucial mind-control technique. "How else," asks Sun Liping, "are people supposed to accept an idea which is quite obviously ridiculous?" Fatiguing tactics break down every last shred of the mind's resistance: you lose all curiosity and all ability to defend yourself, and you allow them to stuff you full of "rubbish." "In the end," writes Sun, "something fundamental is destroyed, namely your will and your ability to throw this rubbish back out."

Meanwhile the Party's central propaganda authority watches over everything. Qiao Mu, a Beijing professor of journalism and one of the few outsiders to have seen inside its offices, tells of a motto displayed on the wall of the conference room:

Responsibility as heavy as a mountain
    As hard-working as an ox
Precise as a hair
    Lips sealed as tight as a bottle
United as one

The Communist Party was, in its own estimation, never anything but "great, glorious, and correct" (*wei guang zheng*). Since the founding of the People's Republic in 1949, the Party has exhorted China's media to be its "throat and tongue" and to make sure that every last subject internalizes the deeper truth of this assertion. Even today, there is still not a single newspaper, website, or radio station that doesn't officially answer to a Party organization. The Party calls its own press the "mainstream media" (*zhuliu meiti*). In this country, "mainstream" is not defined by the people—it's what the Party sets before you.

Censorship and propaganda never rest. Beijing sends out daily directives stating what topics are taboo and which words are forbidden. And yet in the three decades of reform and opening-up, the media landscape was transformed. From the 1980s, Chinese journalism was supposed to start financing itself, just like in the rest of the world. The newspapers had to become more commercial and more professional; a few even dared to print hard-hitting investigative articles. Soon after Xi Jinping was inaugurated as Party leader in November 2012, though, the climate for such journalism became noticeably cooler and the freedoms it had enjoyed were restricted. Overly headstrong reporters and editors-in-chief were fired, and some ended up in jail.

In 2016, shortly after banning the cadres in his own party from having "inappropriate" political discussions, Xi Jinping launched a campaign to completely homogenize the media once again. He visited some of the country's large newspaper offices, bearing a single message: from now on, all media organizations must "have 'Party' as their surname." Without exception, everyone now had to fall in line with "the will, the views, the authority, and the unity of the Communist Party." A subsequent comment piece published by the Xinhua News Agency said that "guiding the public is a great tradition of our Party." "The media has to restore the people's trust in the Party," said *China Daily*, Beijing's English-language propaganda paper for foreigners—especially now, when "the economy is growing more slowly." The Party papers were quite open about why the leadership felt it had to take this action: many people in China strongly mistrusted the CCP and its propaganda. According the *People's Daily*, this was due to the rise of social media, which had opened up a growing divide: "And if this divide remains, it will undermine the legitimacy of the Party's rule." Xi and the Party now demanded one thing above all from every newspaper editor: they had to spread "positive energy."

The process of homogenization has now essentially been completed. Newspapers like the *Southern Weekend*, which were once astonishingly courageous within the limits placed on them, have been brought in line. Many of the dedicated and brave journalists who contributed to the brief blossoming of their craft have either been fired or have given up their jobs. One of them is Luo Changping, until recently the country's most famous investigative journalist, whose articles once even brought down a minister. He was sidelined by his publication, the economics magazine *Caijing*, and eventually left. During his active period at *Caijing*, he had been able to publish between 90 and 100 percent of his material, he said afterward. "Now, they can only publish about 10 percent. And they are the media outlet with the most freedom."[34]

A few years earlier, in 2011, Hu Zhanfan, then head of the state television network CCTV, had warned people like Luo Changping: "A number of journalists no longer see themselves as Party propaganda workers; they have redefined themselves as professionals— but this is a fundamental misunderstanding of their identity." The internet was quickly awash with parodies: "A number of people no longer see themselves as slaves; they have redefined themselves as human beings—this is a fundamental misunderstanding of their identity." It was the time before the web had been completely "harmonized."

As in every other part of the world, the traditional

media in China have long since lost their central place in the news cycle. The Party is now seizing on every channel it can to try and reach young people. Sometimes it works; sometimes it doesn't—though many of the video clips are now strikingly well-made, often with help from professional PR firms. For a while, the "vocaloid" Luo Tianyi was a hit with teenage girls: the virtual singer appeared onstage as a holographic projection and filled huge stadiums. Luo reached millions of fans on social networks, and the Communist Youth League recruited the digital pop star—programmed by a Japanese company—as its "youth ambassador." The Youth League went on to fill her lyrics with "positive values" to "inject young people with correct thinking," as the Party newspapers proclaimed with great pride.

Luo has genuine fans, but other attempts by propagandists to appropriate parts of youth culture have been rather painful to watch. The hip-hop scene popular among China's young people has always been viewed with a degree of mistrust by the state censors; they find it "improper, vulgar, and obscene." According to their censorship decrees, the artists and their lyrics (Sex! Drugs! Justice!) often have "moral failings" and go against "the core values of socialism."[35] So the Party has begun to produce its own rap songs. A few years ago, it started putting out raps to cute animated videos, with titles like "Let Us Look to the Central Committee for Comprehensively Deepening Reforms" (sample lyrics: "The reform group is two years old now / and it has already done quite a lot / Reform! Reform! Reform! Reform!"). Now the next phase has been launched: the Party press is calling it "harmonization of hip-hop," and it is bringing genuine rappers into line. The "CD Rev" (Chengdu Revolution) Crew, for example, have rapped their way into the Party's good books with songs like "This Is China" ("The red dragon is not evil / it is a land of peace"). They have even been allowed to perform to the troops in the South China Sea. Then there's the rapper Sun Baiyi, whose song "Splendid China" begins: "We all know the original vision and the mission of the CCP; it works tirelessly for the happiness of the people and the resurgence of the nation." Politically correct, but so far winning little applause from the target market.

A much cleverer move has been a wave of action films bringing China's new global relevance and the nationalism fueled by the CCP to the big screen as entertainment, using Hollywood methods for the first time. Take *Operation Mekong* and its sequel, *Operation Red Sea*. They feature warlike heroes who are quite obviously modeled on Rambo, but who tie the Chinese flag round their arms at key moments. With fight scenes, explosions, and pursuits through exotic landscapes far outside China's borders, they battle the evil in the world—especially when it lays its hands on Chinese citizens. These films capitalize on the contemporary zeitgeist within the country: China is a powerful nation, its companies and soldiers now operate worldwide, and Chinese heroes help justice to triumph.

In 2017, *Wolf Warrior 2* used this formula to become the most successful Chinese film of all time, giving the Chinese audience, in the words of one observer, "waves of nationalistic orgasms."[36] Its hero, Leng Feng, has gone to Africa to avenge the murder of a woman he loved, but he quickly finds himself duty-bound to rescue a group of Chinese countrymen from a murderous troop of white mercenaries. The film is a paradigm for a new alliance of commerce and propaganda: it was privately financed, but sailed in the slipstream of Party propaganda and was advertised with the PR line: "Anyone who attacks China will be killed, no matter how far away he is."

The film hit cinemas just weeks after the Chinese People's Liberation Army opened its first overseas base in Djibouti, and just a few days before the Army's 90th birthday. It was "a metaphor for the era of China's rise" wrote one cultural studies expert in Beijing, a "collective manifestation of the China dream."

"Perhaps the Chinese have buried their patriotism for too long," said the director and lead actor Wu Jing. "The passion had dried out like wood—and my film is the spark that will reignite it." The final scene shows the image of a Chinese passport, onto which are projected the words: "Citizens of the People's Republic of China! If you encounter dangers overseas, do not despair! A strong motherland* stands behind you."

In the first quarter of 2018, China overtook the USA to become the country with the highest box-office revenue in the world. Well-produced Chinese films increasingly outstripped the Hollywood blockbusters in terms of sales. "Until about 10 years ago, in the absence of a well-articulated set of national and cultural values,

---

* Other translations have preferred the word "nation," but I wanted to convey some characteristics of the Chinese term. The word 祖国 is best rendered as "ancestral land"/"homeland," and is neither male nor female; like "the ancestors" themselves, the ancestral land is an object of veneration in Chinese culture. "Motherland" seems the most convenient translation, because mothers share obvious life-giving and life-sustaining qualities with the land, and because there are similar concepts such as 母亲河 and 母语 in Chinese, meaning "the mother river" and "the mother tongue." However, it would be wrong to suggest that the Chinese view their "motherland" as a female, or to overlook the importance attributed to both ancestors and the ancestral land in China's culture.

Hollywood stories and the ethics they promoted were aspirational to many Chinese," the Shanghai online publication *Sixth Tone* commented. "But now, we're tired of them."[37] The "American Dream," writes *Sixth Tone*, is just a cliché, and Western individualism and liberalism are "simply not seductive to Chinese audiences, who are increasingly embracing a markedly different form of national pride. In China, the ideals of the American Dream are paling into insignificance as the country's self-styled Chinese Dream produces the growing wealth and status of large numbers of people while promulgating different values: collective effort, patriotism, and self-sacrifice for the cause of national rejuvenation."

The English-language website *Sixth Tone* quoted here, and its Chinese-language sister-publication *The Paper*, founded in 2014, are themselves the creation of a professional propaganda machine well aware that its message needs urgent repackaging. They are Party-owned websites, but they have a fresh, modern look, and sometimes push the boundaries of censorship with original, well-researched stories. These stories convey the message—which at its core always affirms the system—much more skillfully than the *People's Daily* and its ilk.

The hard-working oxen in the propaganda apparatus have two principal tasks to carry out. The first is to formulate the Party's message to convey it into the "hearts and minds" of the people. Classic propaganda. The other is censorship: controlling the flow of information to the people and then stemming and channeling the opinions the people express. The Chinese are even fonder of hydraulic metaphors than my fellow Germans. "To stop the mouths of the people is more dangerous than damming a river," goes a saying that originated over 2,500 years ago.

When *The Paper* interviewed the director of the Party School's magazine *Seeking Truth*, he likened this control of public opinion to the mission of the Great Yu, the mythical ancient emperor who tamed the huge river in the Chinese legend. The Great Yu had learned from the catastrophic experiences of his predecessors, who could come up with no better solution than simply damming the river, though the dams soon broke again. "But the Great Yu relied on a combination of damming and channeling; he stopped the flow in one place and let it go again in another. That brought him success. And this is even more necessary with public opinion."[38] Free speech is a natural force that requires clever taming.

In the old world, that was a simple matter. Then the internet arrived, and suddenly even ordinary citizens had a chance to be heard—by thousands, hundreds of thousands, millions of people.

The Net

How the Party Learned to Love the Internet

"The old and the new conspire to shut us in.
When will we stop adding new stones to the wall?
The Great Wall of China: a wonder and a curse!"

Lu Xun, 1925

New forms of media always promise to empower the powerless, and from the outset, this was also the dream of the internet. But such dreams always threaten the status quo. China's attempt to censor the web, as the former US president Bill Clinton joked, was like "trying to nail Jell-O to the wall." That was in the year 2000. The Chinese listened to the prophecy, and swiftly built a new great wall: the Great Firewall. They banged a few nails into the cracks, and look: the jelly is staying put.

The prophets of freedom didn't let that discourage them. In November 2013, Eric Schmidt added his voice to the chorus of stubborn optimism. It was a foregone conclusion, a race between the hare of censorship and the tortoise of netizens: "First they try to block you; second, they try to infiltrate you; and third, you win." The then-chairman of Google predicted the defeat of web censors all over the world within a decade. Speaking in the same year, Tim Berners-Lee, inventor of the World Wide Web, said: "The Berlin Wall tumbled down, the great firewall of China—I don't think it will tumble down, I think it will be released." Piece by piece, website by website. "The agility of a country which allows full access to the web is just greater; it will be a stronger country economically as well."

For a while the net optimists were able to bolster their confidence with stories of the Twitter uprising in Tunisia, the Facebook revolution in Egypt, the YouTube agitation of the Maidan activists in Kiev. And then there was Hong Kong, once in the summer of 2014 and then again in 2019, where the people, especially the youth, occupied the streets of the city and organized themselves through Firechat, Telegram, and the LIHKG forum (the "Hong Kong Reddit"). The vocabulary of rebellion—especially among awestruck Westerners—was augmented by the app arsenal used by cleverly organized students. And, of course, for the media, the fight for freedom and its technological foundations often go together.

It may seem surprising, then, that Beijing doesn't seem to fear the net at all. More than 900 million Chinese people currently use the internet, the overwhelming majority of them via their smartphones. And the government is working full tilt to extend the infrastructure. The Hangzhou online marketplace Alibaba reports higher revenue and profit figures than Amazon and eBay put together, and it now operates in more than 200 countries worldwide. In November 2017, the Shenzhen internet giant Tencent overtook Facebook as the world's highest-valued social-network company. Both now number among the ten largest and most valuable companies in the world. Party leader Xi Jinping has celebrated China as a "cyber-power," proclaiming: "Let us climb aboard the express train of the internet." The Xinhua News Agency speaks in glowing terms of China's "power of innovation," which is thanks to a globally "unique method of managing the internet"— commerce and control united to extremely fruitful effect.

The World Internet Conference was initiated by

China and has taken place annually since 2014 in Wu-zhen, in the country's south. In December 2017, it greeted delegates in almost euphoric fashion: "China has [. . .] countless internet gurus," writes Xinhua, "but none can measure up to the 'wise man of the internet' Xi Jinping."[39] Xi is not only transforming China into an "internet superpower," he is simultaneously making sure that everyone can sleep soundly again. "Should we simply invite burglars into our houses? Of course not! And what do we do when they arrive at the door uninvited? We build high walls and use large padlocks. What we do with our homes, we must of course do with the great, wide internet."

There is no doubt that China loves the net. And the Party *really* loves it. Until quite recently, though, even in China itself more than a few people shared the optimism of the West's net prophets. Liu Xiaobo, the late Nobel peace laureate, called the internet "God's gift for a democratic China," and the artist Ai Weiwei prophesied a fatal blow for censorship: "The people will always have the last word—even if someone has a very weak, quiet voice. Such power will collapse because of a whisper," he wrote in a euphoric essay. "The internet cannot be controlled. And if it is uncontrollable, freedom will win. It is that simple."

Ai Weiwei wrote these words in 2012, shortly before

the Party showed the Chinese people and the world at large that, yes, it was very simple—only not in the way most people had imagined. The following year Xi Jinping took office, and he lost no time in showing the world how one goes about taming the internet.

The four years leading up to 2013 may well lodge in Chinese popular memory as the years when, for the first time, people felt that they belonged to a society that deserves the name. When citizens learned to pool their knowledge; to connect with each other; to have public discussions about the things that determined their lives. All unheard-of phenomena in a country where the Party had always sought, if not to forbid, then to collectivize and supervise all cooperation, down to the smallest village, the smallest business, the smallest club. For this Party, words like "society" and "the people" had been merely useful weapons with which to castigate shameless foreign countries: for instance, when they had "hurt the feelings of the Chinese people."

But now society had awoken, and it was escaping the confines of the Party's narrative. The free word had fallen into citizens' laps, and the result was a spontaneously growing network of independent-minded citizens. Thanks to the internet. More precisely: thanks to social media, and in particular Weibo. Sina Weibo went online on August 14, 2009, as China's answer to the American platform Twitter—which the Party had banned as soon as it emerged, along with Facebook. During those four years, Weibo acquired a vast importance in this country where traditionally all media had been firmly in the Party's grip. A good way to tell this story is using the case of the blogger and "word criminal" Murong Xuecun. He is the writer to whom, in the previous chapter, a school-friend, now a Party functionary, had explained the message behind the walls of propaganda: "We can cover the walls with this stuff. Can you?"

Just a few months prior to that conversation, Murong Xuecun could have replied: "Yes, I certainly can!" At that time one of his Weibo blogs had four million followers. This was before the deletion brigade arrived and gave him a good talking-to. Before he became the first major blogger in China to be forced back into darkness and silence. You can see why he was the first one they tackled. Murong Xuecun, born in 1974, is one of the country's most sharp-eyed social critics—certainly one of the most sharp-tongued—and he had Weibo to thank for allowing him to become these things. It was the internet and social media that first gave the country the figure of the public intellectual, who could spark debates before an audience of millions.

This was the second online rebirth of Murong Xue-cun. The first had occurred in 2002, when he was working as a freshly-appointed executive at a cosmetics company. He had been sitting in his office, bored, surfing the internet, when he discovered that other people were posting little novels they had written there, untroubled by paranoid editors and over-zealous censors. *I can do that too*, thought Murong Xuecun, whose name then was still Hao Qun; *and I can do it better!*

So he began to write about the tragicomic experiences of three young men trying to make their way in Chengdu, a story of drinking and gambling and bribery and betrayal and whoring in modern China. He posted it online one chapter at a time, and eventually accumulated five million readers, got his first book contract, won his first literary prize, and was chosen as "personality of the year" by the glossy magazine *New Week*. The internet had made the 28-year-old a literary star. Murong Xuecun was the figurehead of a development that turned China's literary landscape upside down: all the new discoveries, all the literary trends of that decade came from the internet. Suddenly readers were hearing voices they'd never heard before.

Murong Xuecun gave up his day job and carried on writing wild, comic, bleak novels about the excesses of the new China, about greed, sex, and violence, and he went on being successful. But the reality of the non-virtual China also quickly caught up with him. This was a realm where the old autocrats and bureaucrats still held sway. In China, as in other countries, the way writers earn money is still with the traditional publishing houses, who fish authors out of the net and then straighten them out: *enfants terribles* with Chinese characteristics.

Murong Xuecun started breaking off works-in-progress because he knew they wouldn't get past the censors. Things finally came to a head when he had spent three weeks conducting an undercover investigation of a fraudulent pyramid scheme in Jiangxi Province and produced a reportage book out of it, *China: In the Absence of a Remedy*, which was only published in Chinese. The book won him the high-profile People's Literature Prize in 2010, but interventions by his editor had made him so angry that his acceptance speech was an act of reckoning with the Chinese censor. "It stuck in my throat like a chicken bone, I just had to spit it out," he told me. The propaganda minister was in the audience. Just before the author went on stage, his editor begged him not to read out the speech. And when he went out to receive the prize, one of the people on stage with him actually blocked his way to the speaker's lectern. "I really thought that was ridiculous.

I'm actually quite an affable character. But I don't bow and scrape to anyone." The prize winner turned to the audience and moved two fingers slowly across his lips, as if zipping them shut. The gesture instantly increased his fame still further.

Afterward he posted the speech online, in full. "I don't want to call myself a writer," it said. "I'm just a word criminal." Ultimately there was only one truth in China, the speech went on, and that was: "We are not allowed to speak the truth." Murong Xuecun spoke of the shame of self-censorship. "Why isn't China producing any great writers?" Because they are all, himself included, "castrated eunuchs," castrated by their own hand in over-hasty obedience, "before the surgeon can even raise the scalpel."

The third life of Murong Xuecun, as a political essayist and public intellectual, began at that moment. "It wasn't an easy decision," he says. "In this country, it's risky." Especially when you think things through to their logical conclusion, as the blogger Murong Xuecun did from the beginning; when you don't practice the art of beating about the bush like everyone else. Like other critical voices, he set up one, two, three, four accounts on Weibo: when one was blocked, he quickly moved on to the next. He became part of the emerging Weibo Revolution. He wrote about power without limits: "A monster on the loose." About corruption: "In this country, it's the norm, the unwritten law. You don't need to get involved in corruption, it gets involved in you. It follows you everywhere. No one stays clean." About the system: "For six decades now, the Chinese have been living in a system that dulls their minds and makes them hate each other. In this system everyone is a criminal, so no one ever needs to regret anything."

Astonishing things happened over those years. Isolating your subjects is one of the oldest power techniques in authoritarian systems. "No fraternization" was Aristotle's imperative for tyrants. As he argues in his *Politics*, the tyrant must "stifle everything that might give rise to two things, self-assurance and trust." Citizens must remain estranged from each other, "for if they become properly acquainted, they begin to trust one another." In China, this isolation suddenly broke down. It was something the traditional internet had never succeeded in doing: an army of censors ensured that the Party had always had a good grip on it. The Weibo headquarters, and the offices of the other new social media firms—all of which had come from the private sector, and not from the old propaganda apparatus—had been staffed with thousands of censors from the outset, on the authorities' instructions.

At first, though, the Party didn't understand what

was fundamentally new about social networks. People were no longer sitting at home in front of their computers; gadget-mad China was even ahead of Europe when it came to buying smartphones. Everyone had one, and they were using them to access the internet. You could now send a message or a photo from your phone, and it would reach hundreds of thousands or even millions of people in seconds. A censor who took just twenty minutes to find and delete something was twenty minutes too late. And so in those four years, Weibo became a realm of unprecedented freedom, especially for China's young, urban generation.

At that point, Weibo officially had 350 million users, 50 million of them active on a daily basis. Food scandals, pollution, police violence—all at once, Chinese people had access to information they had previously been denied. Users forced the city of Beijing to publish its air pollution levels. They shared jokes, Hello Kitty cartoons, and conspiracy theories, just the same as people all over the world, but they also debated China's constitution, the tirades of Maoists, and essays by liberal celebrity bloggers like Murong Xuecun or Han Han, who at the height of his fame had 50 million followers, before growing steadily more cautious in tone.

"With Weibo, people started thinking about things,"

says Murong Xuecun. "There was an awakening: individual, political, aesthetic, cultural." Once you were allowed to speak and think freely, the author believed, you would quickly dig up the treasure that had been buried deep under the Party's garbage thoughts and garbage words: common sense. "Like all authoritarian systems, this one depends on the fact that you are a lone person faced with an overpowering organization, and you capitulate. Since Weibo arrived, that no longer works. People are forming networks." Suddenly, China was a different country. For the first time since the People's Republic was founded in 1949, there was a public space that belonged to citizens, where their language was spoken. The germ of a civil society began to grow.

This alone would have been enough to unsettle the Party. But there was a more immediate challenge: one of the most spectacular phenomena of the new net was the "human-flesh search engines" (*renrou sousuo*). Users cooperated in their hundreds of thousands to gather information and data about particular people and their alleged misdemeanors. This wasn't always a blessing; it frequently degenerated into smear campaigns against innocent people and into anonymous trolling of "traitors to the motherland" and other ideologically unpopular people. But in 2012, with the entire country's approval, the online swarm blew the horn to start the hunt for corrupt officials and Party cadres.

The most prominent case was 56-year-old Yang Dacai, a functionary from Shaanxi Province. There were photos of him circulating online that at first glance looked like standard-issue propaganda: a well-fed Party cadre on a tour of inspection, serving the people, posing for photographers on building sites and in conference rooms. But perceptive netizens had noticed an interesting detail: Mr. Yang's watches were of the finest Swiss manufacture. They identified a Rado, a Rolex, two Omegas, and a Vecheron Constantin—total value: around $100,000. Far beyond what a provincial functionary could afford on his regular salary. From then on, Yang was the "Watch Uncle," and the Party had another scandal on its hands, reinforcing the impression among the Chinese people of a CCP that was now ravaged by legendary levels of corruption.

More important, the citizens had discovered their own power, and the Party was unnerved by it. Would all those prophecies that the new media would help empower the people, bring freedom, and subvert the authoritarian regime be proved correct after all?

The new man at the helm gave the answer to that, starting in late 2012. And it all happened very fast. Party newspapers identified the internet as an "ideological battlefield" where "the hostile forces of the West" were running riot: "He who wins the battle of the internet will win the war," wrote the Shanghai CCP paper *Liberation*. In August 2012, Xi Jinping gave the order to "win back the commanding heights of the internet." And in November a deputy propaganda minister reported that the mission had been accomplished: "Our internet is clean again." The scare was over. Observers on the perimeter of the battlefield rubbed their eyes: was it really that easy?

It was that easy. The Party had brought out its old weapons: intimidation, censorship, propaganda. Polished up and cleverly adapted to the times. The initial offensive was led by intimidation: first, the Party deleted the accounts of bloggers that were making it feel uncomfortable; Murong Xuecun was the first. When his accounts were deleted, and four million readers were wiped out at the click of a button, he felt as if he were "being pushed back into isolation." The spreading of fear is the autocrat's core competency. Faced with a machine that can squash an individual—and his family—with a snap of the fingers, only a few people will ever possess the heroic courage required to fight back. The reconquering of the internet in the summer of 2013—and beyond that, the successful exploitation of the internet to promote the aims of the regime—

demonstrated how efficient the system was once it had identified the enemy. It might be uncharted territory, but the state apparatus still worked.

Murong Xuecun's account was deleted in May 2013. The real campaign began on August 10. On that day, China's foremost opinion-leaders on the net, the so-called "Big Vs" (the V stands for "verified account"), were invited to a conference in a Beijing hotel. Many of the Big Vs were prominent entrepreneurs; others were film stars or singers; most had made a name for themselves as liberal voices. The conference was hosted by the head of a new organization currently in its start-up phase, the Cyberspace Administration of China (CAC). At the event, these prominent bloggers were reminded of their "social responsibility," and that each of them had a duty to promote the "interests of the state" and "core socialist values." The conference proceeded in a more or less friendly atmosphere; the first painful blow came two weeks later, on August 23.

On that day the people of China learned that one of the Big Vs, the venture capital investor Charles Xue, had been arrested. Xue had 12 million followers on Weibo, and he had used this platform, among other things, to call on the government to devote more attention to clean air, healthy food, and the problem of child abduction. TV viewers were able to watch the police raiding his apartment late at night, complete with flashing blue lights and a lot of shouting, before taking him and several women away. The state media accused him of "encouraging prostitution" and "pimping"—not forgetting to mention in the same breath that Charles Xue was a notorious "agitator" and "spreader of rumors" on Weibo. And then Charles Xue disappeared—only to turn up again a short while later on state television, behind bars and wearing prison garb.

It was one of CCTV's first "interviews" with repentant sinners, of the kind that have since become a regular feature of the viewing schedule. With tears in his eyes, Xue confessed how "irresponsible" he and other bloggers had been to express their opinions freely on Weibo without guidance from the Party, simply "to satisfy my vanity," and how right the Party was to put a stop to this irresponsible behavior.

Blow number three finally came from the Supreme Court. In September 2013 it issued a new ruling: from then on, anyone who spread a "rumor" that was shared more than 500 times or received more than 5,000 clicks and thus "upset social order" would risk up to three years in prison.

500 shares? 5,000 readers? Prison? You can imagine the shudder that ran through the celebrity bloggers with their two, five, or ten million followers when they

read this ruling. They fell silent. All of them. Ever since, Weibo has been dead as a politically relevant medium. Once, debate had raged there: sometimes wild, often polemical, clever if you were lucky—but always lively. Today, it's as silent as the grave.

But wait—Weibo still exists. It's more commercially successful than ever. Commerce and entertainment are its stock in trade; you'll find glamour and glitz, celebrity gossip and propaganda, and the site is peppered with posts by various CCP organs and people writing covertly for the Party, who are known as the "fifty-cent" (*wu mao*) troop, because once they were supposedly paid half a Yuan for every online comment.

Shortly after Charles Xue's enforced TV appearance, I visited Huang Chuxin in Beijing. He is one of the authors of *Blue Book on Development of New Media*, which the Chinese Academy of Social Sciences (CASS) had just brought out. Huang was working as a researcher for CASS, and at the same time as a social media consultant for various government publications. In the week that I met him, he had already been to Hunan, Shandong, and Inner Mongolia to tutor local government officials in how to deal with Weibo and other sites.

Huang spoke in euphoric terms about the potential of the new medium: "The opportunities for the government here are huge," he said, "and certainly greater than the dangers." The central government, said Huang, had finally realized that—with a few skillful modifications—social media could provide new channels through which to gauge the mood of the nation. At the same time, the Party was now encouraging civil servants and government agencies to open their own Weibo accounts *en masse*. "They need to have conversations with the people. It makes politics more transparent and the apparatus of government more effective." Of course, these functionaries now had to learn to react quickly and use ordinary language. "The *People's Daily* already does that very well," said Huang. Unfortunately, among the civil servants on the ground, his overriding impression was that they were "afraid of Weibo." When he was teaching his courses, he told me, the first question people asked was always about crisis management. "How do we get a crisis of public opinion under control?" By which they meant criticism from the people.

The first test of strength for the world's greatest censorship and propaganda machine came just a few months after it had reconquered the "commanding heights of the internet." The former British Crown colony of Hong Kong had been returned to China in 1997, when Beijing pledged to uphold for 50 more years the city's new "Basic Law," which guaranteed autonomy and the maintenance of Hong Kong's freedoms. Hong Kong's citizens protested Beijing's increasing interference and the creeping erosion of democratic freedoms in their city, demanding that their prime minister be directly elected, which they were promised in Basic Law. Because the demonstrators put up umbrellas against the tear gas of the police, the movement was soon called the "umbrella revolt." How did the Party respond? It did not give in to a single demand of the demonstrators. And in the end they retreated. Censorship works.

In the summer of 2014, protests by citizens in Hong Kong brought the city to a standstill. Right at the start of the protests, the censors shut down the photo and video platform Instagram. Hong Kong users of the popular messaging service WeChat noticed that although they were still able to send messages, the intended recipients in mainland China weren't getting them. The Hong Kong–based anti-censorship project Weiboscope reported that the day after the police used tear gas on the demonstrators, a record number of messages on Weibo were deleted and blocked—more than on any other day that year. The news blockade was effective from day one. Hardly anyone in China found out what was really happening in Hong Kong, not in 2014, when its citizens were on the street, still hoping, and not in 2019, when hope had already died and, above all, driven them to despair. The machine had performed splendidly.

Bao Pu is a publisher in Hong Kong, but he grew up in Beijing, as the son of a senior Party official at the heart of power. He is familiar both with the CCP and with the new technologies. For that reason, he says, he's a pessimist. "Technology," says Bao Pu, "always benefits the side with the greater resources. So the internet will always serve the CCP more than its opponents." The all-powerful secret police now just have to read Weibo and WeChat, he explains, "and they know who to arrest next."

In the wake of the censorship of Weibo, the newer WeChat quickly overtook the older service in popularity. But WeChat is fundamentally different. Weibo was a megaphone: you could reach thousands, hundreds of thousands, millions of people with it. On WeChat, people connect in small friendship groups. The maximum number of people allowed in a group is 500. People just talk to their friends. The exodus from Weibo to WeChat was a retreat into the semi-private sphere. The censors appear to be more tolerant on WeChat, but the feeling of relative safety that many users have there is deceptive: after planning on WeChat to attend a poetry reading in support of the Hong Kong demonstrators, a group of Beijing residents found themselves arrested as they met up in the Songzhuang artists' quarter. The state security service reads your messages.

The protests in Hong Kong were an object lesson in the new style of internet management, in which the aim is not only to block information. The Party immediately provided its own counter-narrative. Propaganda has become more sophisticated. Yes, the old-school rabble-rousers were still there, foaming at the mouth, fulminating in the state press that the demonstrators had left behind "a stink that would last ten thousand years." Yet far more effective were the other pieces, the professionally produced infographics, the anonymous blogs disguised as contributions to the debate from concerned citizens, explaining the situation to their countrymen: Hong Kong is descending into chaos. The demonstrators want to split Hong Kong off from China. They're spoiled children, who only care about their own economic advantage. They're backed by shadowy foreign powers—"black hands"—who don't want China to become strong.

It was also an exemplary demonstration from the Chinese propaganda machine of how to hijack the terminology of the West. The *People's Daily* even wrote that the problem with the 2014 demonstrators in Hong Kong was their "anti-democratic" attitude. This "hostility to democracy" had been sown in their minds, so the Party propaganda said, by British colonial rule. Now the first priority was to re-establish the "rule of law" in the city—which is to say, crush the demonstrators once and for all.

At the same time, all words that might serve the regime's opponents were blocked online. You couldn't talk about "umbrellas" on Weibo anymore. Users were creative, of course, and took a mischievous delight in sneaking the characters for "Pearl of the Orient" past the censors once "Hong Kong" had been blocked— but Beijing had no need to worry about these little triumphs. The language that was central to the discourse of the demonstrators and their sympathizers had been taken from them, and with it the prospect of having any substantial impact. The word, the gun, and the pen came together online to form an invincible weapon for the Party.

And in 2019 CCTV even stole the famous Holocaust poem by the German Protestant pastor Martin Niemöller, in which he describes the failure of his fellow travelers in the face of a criminal regime:

First they came for the Communists
And I did not speak out
Because I was not a Communist
Then they came for the Socialists
And I did not speak out
Because I was not a Socialist
Then they came for the trade unionists
And I did not speak out
Because I was not a trade unionist
Then they came for the Jews
And I did not speak out
Because I was not a Jew
Then they came for me
And there was no one left
To speak out for me

In their version of the poem, the CCTV propagandists put the protesting students of Hong Kong in place of the Nazis:

When they blocked the streets and arrested
    and abused drivers
I was silent
I was not a driver
When they pushed and attacked passengers

I kept silent . . .

I was not a passenger.

The CCTV verses ended with the words: "When they attacked me, there was no one left to speak and protest for me."

China's propaganda had already started to call the demonstrators a "violent mob" and their protests an "insurrection" that carried "trains of terrorism." Hong Kong head of government Carrie Lam spoke of "enemies of the people," and police officers of "cockroaches," using old communist rhetoric.

Hong Kong in 2019 was also the first time Chinese propaganda on a large scale orchestrated a disinformation campaign on Western social media channels: Twitter and Facebook announced in August 2019 that they had closed several hundred accounts that were active for the campaign. Without mentioning China, Twitter spoke of an "organized operation with state support."[40]

This episode showed what an ingenious move it is to steal your opponents' core terminology and give it the opposite meaning. In the case of the Hong Kong protest, the astonishing thing was not the show put on by the propaganda machine, but how obediently the people played along. While the mighty army of state-sponsored "fifty-cent" trolls castigated the citizens of

Hong Kong online as ungrateful so-and-sos with no love for their great nation, you could hear the echo of this propaganda in the real world, in everyday conversations on Beijing's streets, in exchanges with acquaintances.

"Hardly any of my friends know what's really going on here," said Murong Xuecun, whom I bumped into on the periphery of the demonstrations in Hong Kong. "And of those who do know, 70 percent are complaining about the Hong Kongers. They just can't imagine anyone standing up for an ideal these days, because in China all idealism, all principles, all morality has been wiped out. People only work for their own self-interest and their own profit, so they always see the lowest motive in everyone else, too." In China, there is a saying for that: "To look at humans with the eyes of a pig." In 2019, the propaganda message that the Communist Party delivered to the people on the Mainland was that Hong Kong was sinking into chaos and violence and its economy suffering. For many years the Party has fueled the fear of "color revolutions" and foreign powers that sow chaos, violence, and decline around the world in the name of democracy. "The desire for western liberal democracy is an evil virus that infects places with weakened immune systems," warned *China Daily*. "Countries like China, Russia, Iran, and

Cuba have demonstrated the ability to resist the disease and deterioration that such an infection often leads to." Hong Kong, however, remains weak: "A weakened immune system will always lead to future infections." In the end, the paper concluded, the best cure for the residents of Hong Kong was the same as that for those other elements who have fallen victim to terrorism—an obvious allusion to the Uighurs in Xinjiang: "Education." Re-education, reformation, brainwashing. "This is the best way to respect and protect human rights," a Chinese diplomat is quoted as saying.[41]

A vast apparatus takes care of the day-to-day business of censorship. In 2017, the censors blocked 128,000 websites for containing "pornographic and vulgar information" and exerting "a negative influence on public opinion online." The security apparatus regularly practices an "instant switch-off"—in August 2017, for example, two months before the last CCP Party congress. Over the course of an exercise lasting several hours, internet hub administrators have to prove to the Ministry for Public Safety that they are capable of taking down a series of specific websites in a matter of minutes. Following an outbreak of unrest in 2009, technicians took large parts of the Uighur province of Xinjiang offline completely for a considerable time. Since the Party ordered China's internet engineers to build a "Golden Shield" in 2003—an "internet with Chinese characteristics"—they have achieved incredible things. China has built a net using hardware and software completely independent from the rest of the world. Today, China's net is more intranet than internet. If the censors were to cut off all links to the outside world overnight, most people wouldn't even notice.

Regular updates of the list of words banned on Weibo and other platforms have become part of everyday online reality. At the start of 2018, when the change to the constitution that allowed Xi Jinping to be president for life became public knowledge, that list was particularly long. There was no place on Weibo for obvious words like "accession," "my emperor," "all hail the emperor"—nor could you make reference to *Animal Farm*, George Orwell's parable of totalitarianism. Winnie the Pooh was banned in both text and image; in recent years, China's online satirists have turned the Disney version of the tubby bear into a *doppelganger* for the Party leader and head of state. For a while, people weren't even allowed to post the name Xi Jinping. More remarkable still was the block on the three characters *bu tongyi*. They mean "don't agree." Anyone trying to type the words received an error message and was informed that unfortunately, the statement "I do not agree" "violated laws and regulations."

The change in the constitution was seen as a historic moment by many people, and as a break with the legacy of Deng Xiaoping, under whose rule the limits to terms of office had been introduced in 1982 in response to the tyranny of Mao Zedong. Many intellectuals and members of the urban middle classes were shocked by Xi Jinping's move. During the meeting of the National People's Congress, which passed the amendment in March 2018 with just two votes against and three abstentions, the altered constitution was the number one topic for many Chinese citizens, and for the foreign media. But on the day of the vote, the news didn't even make the list of most-read and most-discussed topics in China's sanitized social media world. Instead, in the hours after the decision, Weibo's top ten included an article entitled "When Your Girlfriend Has Loads of Money." Also near the top of the list was the crucial debate: "Is it okay to eat instant noodles on the high-speed trains?"

An acquaintance who had moved to Berlin wrote on Facebook about one of her Beijing friends, an editor so frustrated by the automated online censorship that he'd spent the whole night dreaming that he was desperately changing every potentially sensitive word in an article over and over again in an attempt to get past the censor. "It reminded me of a cat who smells fish in a dream and starts running, even though she's lying on her side, fast asleep," wrote my acquaintance, who had once been a journalist in Beijing herself. "A great deal of trauma, every day."

The internet has changed the form that censorship takes. The state apparatus no longer has to worry about the traditional Party media; it has turned its attention to the start-ups and the internet giants, almost all of which are private-sector companies. Of course, every private company in China knows that it exists only by the grace of the Party, and of course they all praise the CCP and pledge their faithful allegiance to its leader, and of course they have voluntarily placed hundreds if not thousands of censors in their editorial teams from the start. And yet their business model was new to China: fierce competition among countless competitors, the unrestrained desire for profit, boundless commercialization, and the breathless rhythm of continual renewal characteristic of the global online sector. Disagreeable things inevitably slip through the fast-moving net, undetected by the censors, who must therefore constantly refine their methods.

One example that particularly delighted the online community was a pair of chatbots named BabyQ and Xiaobing. Chatbots are systems with which you can converse via text or speech. Their communication is fully automated, with no human input required. These AI-powered creations were designed to chat with users on the "QQ" messaging platform owned by the company Tencent. People soon began to circulate screenshots showing BabyQ answering the question: "Do you love the CCP?" with a simple "No." The bot responded to a cry of "Long live the Communist Party!" with a tirade about the "corrupt and incompetent" system. And if you asked them about the "China dream," one of the bots replied that it would like to emigrate to the USA, while the other said: "The China dream is a daydream and a nightmare."

Tencent's experience with the uncontrollable nature of self-teaching systems was similar to that of Microsoft, whose Twitter chatbot Tay quickly descended into spewing racist and sexist bile. At the same time, there was a flash of realization that the "mainstream" in China might, in reality, be different from the image stubbornly propagated by the Party. The subversive bots were instantly switched off, but those learning algorithms must have absorbed their resentment of the Party from somewhere.

China has a whole set of censorship authorities devoted to the task of giving the country a "civilized net" and making "the Party's positions into the strongest voices on the internet." The Cyberspace Administration of China (CAC) has made it its mission to "synchronize the opinion ecology of the internet with the reality of Party and nation," according to the *People's Daily*. The year after it was founded, the CAC even brought out an anthem for the 2015 Spring Festival (Chinese New Year, as it's known in the West). In a video posted online, a choir of festively-dressed CAC employees celebrated the work of the net censor in song: "Day after day with devotion we watch over this place / An internet power: where the net is, there too is the glorious dream." After the glorious net replied to the song with a shower of ridicule, the censors swiftly censored their own hymn of praise to censorship, and the video disappeared.

Recently, one wave of censorship after another has rolled over the big social media firms and platforms, which are repeatedly accused of spreading "vulgar" and "damaging" content and "rumors." The largest Chinese social media channels, Weibo, WeChat, and Baidu Tieba, have all regularly been placed under supervision for short periods. Each time, these firms are instructed to "clean up their act and correct their mistakes immediately"; they respond with gestures of submission and avowals of their "core socialist values." Sina Weibo, Baidu, and Tencent vie with one another to employ the greatest number of new censors. Tencent has come up with the cute name "Penguin Patrol" for its army of censors (the firm's mascot is a penguin), and enlists not only trained editors but also normal internet users, who are rewarded for hunting down disagreeable content with vouchers and gadgets. New censors are in such demand that the companies have long been forced to advertise publicly to fill the positions.

With 120 million active users every day, Toutiao is probably the most successful news app in the world. In September 2017, the *People's Daily* mounted a fierce attack on the company, accusing it of spreading "uncivilized content" and "sensationalized news." The app was suffering from an "incurable disease," the paper said.[42] The Party's discomfort had been caused by an algorithm deployed by users to filter out political propaganda in favor of more sports or celebrities or scantily-clad women. Toutiao immediately reinforced its team of 4,000 news moderators with another 2,000 staff. The job ad requested applicants with "political sensibilities," who must be capable of scanning up to 1,000 posts a day for illegal content.

Shortly afterward, the company announced that it was extending its censorship team to 10,000 people. And at the start of 2018, after fresh criticism by the censorship authorities, Toutiao blocked 1,100 bloggers, and even briefly replaced the "Society" category on its homepage with one labeled "New Era," featuring articles singing the praises of Xi Jinping's policies. (This, after all, is the most pressing task of "society" in the Party state.) All this, though, still wasn't enough for the censors.

In April 2018, the pressure on the company—then valued at $20 billion—was so great that the founder and CEO, Zhang Yiming, made a public apology. His letter stood out from a crowd of similar declarations for its potent mixture of breath-taking obsequiousness and evident fear for the company's survival, and illustrated just how fine the line is that China's new companies have to walk.

"I earnestly apologize to regulatory authorities, and to our users and colleagues. Since receiving the notice yesterday from regulatory authorities, I have been filled with remorse and guilt, entirely unable to sleep," it begins.[43] Zhang goes on to confess that the quick development of his company has only been possible thanks to the "great era" ushered in by Party leader Xi Jinping: "I thank this era." And finally, he confesses his sins: "I profoundly reflect on the fact that a deep-level cause of the recent problems in my company is: a weak [understanding and implementation of] the 'four consciousnesses' [of Xi Jinping]; deficiencies in education on the socialist core values; and deviation from public opinion guidance. All along, we have placed excessive emphasis on the role of technology, and we have not acknowledged that technology must be led by the socialist core value system, broadcasting positive energy, suiting the demands of the era."

The Party has no fundamental objection to algorithms; in themselves, they are the miracle weapon of the future. A comment piece in the *People's Daily*, however, hoped for a "melding of man and machine" in the interests of a politically correct solution. What the writer really meant was the melding of man, machine, and Party: the red algorithm.

The Party dams the river, but it takes care not to stop it completely. It serves its purposes to give different voices a little room, though never enough for them to become dangerous. In the decades of reform and opening-up, Chinese society fanned out to create a universe of diverse interests and voices. Voices upon which the Party wants to eavesdrop. Even now, debates on social issues sometimes manage to become nationally relevant (if only partially, or briefly), as long as they don't appear to pose an existential threat to the system. The #MeToo debate is one such example.

In spring 2018, millions of people—especially students—were discussing a story from twenty years previously, of a young female student at Peking University who committed suicide, allegedly after being raped by her professor. The more commercially-oriented Party media often draw from social media debates of this kind for their own articles, and sometimes these actually have consequences: Peking University published a statement pledging to do more to combat sexual harassment, and other universities followed suit. But all the while, you could sense the system's usual nervousness about any debate it isn't steering itself—and a movement in which women demand accountability from teachers, professors, managers, and senior officials is bound to cause a backlash in a system like China's, dominated by powerful patriarchs. Here, it is the custom for unassailable men to cover for each other.

Alongside the recent feminist awakening within society at large, a reactionary counter-movement has arisen. This is often backed by the CCP's women's organizations, whose aim sometimes seems to be to cajole women back into submissiveness and other classical feminine virtues.[44] The hand of the censors is never far away. Feminist activists at Peking University were warned in a university WeChat group that their struggle had been identified as a "political movement." If they carried on making a fuss, then before long they might be exposed as traitors who were colluding with foreign powers.[45]

A student named Yue Xin, who had written an open letter about what had happened on WeChat, was taken from her dorm room in the middle of the night and locked up in her mother's house for several days. The incident caused so much outrage among her fellow students that a few anonymous people pinned a wall poster* up on campus and called for Yue Xin's release—there had not been wall posters at the university since 1989. The US-based *China Digital Times* published an instruction from the censors banning all media and platforms from sharing "content expressing so-called solidarity."[46] It is the autocrat's most deep-seated fear: solidarity and cooperation among his subjects. These things have to be nipped in the bud. Activists at other universities in Beijing were warned that their campaigns were unacceptable because they were "planned" and "organized." In China, planning

---

* The Chinese term—"big character poster"—is more precise, referring to the size of the lettering used on these principally hand-painted, visual protests. They have a pedigree stretching back, beyond the brief flowering of freedom and debate in 1978–79 known as the Democracy Wall Movement (also called the Beijing Spring), to the dawn of the Cultural Revolution in 1966, when Mao Zedong himself created one, urging students to "Bombard the Headquarters."

and organizing anything at all are activities reserved for the Party.

The idea of the internet and social media as a feedback loop and a kind of early-warning system for the Party and the government has led to the creation of data-gathering organizations in practically every branch of government and the Party. The Beijing communications researcher Hu Yong, who knows the internet in China better than most, calls it an "industry of public intelligence services." According to Hu, this industry is constantly frustrating its own primary purpose: to sound out the current mood among the population. Its work automatically "results in the contamination of public opinion": in other words, as soon as it detects a deviation from the mainstream, the old reflexes kick in and the authorities "intervene at once to suppress any negative opinion." What begins as an objective data-gathering exercise rapidly becomes "the manipulation of public opinion," says Hu Yong. The security apparatus often stymies itself.

In 2017 a joint investigation by Harvard, Stanford, and San Diego Universities, based on leaked documents from Jiangxi Province, claimed it could prove the existence of a "huge secret operation" serving the Chinese government.[47] According to the authors, in this year alone 488 million government-sponsored comments were posted on China's social media platforms by "fifty-cent" troops. The study revealed that these paid trolls were not just hobbyists, but civil servants specifically assigned to the task. Another revelation was that these trolls generally avoided crowbarring the official government position into controversial debates. Instead, their main techniques were "cheer-leading" for the Party in general ("[If] everyone can live good lives, then the China Dream will be realized!"[48]) and simple distraction: when a sensitive debate had reached critical levels of intensity, they would flood the forums and comment threads with posts that all had a single aim: to change the subject.

A wide range of competing ideologies continues to circulate on the Chinese internet, despite the blows struck by the censors: Maoists, the new Left, patriots, fanatical nationalists, traditionalists, humanists, liberals, democrats, neo-liberals, fans of the USA, and various others are launching debates on forums like Tianya, Maoyan, Tiexue, and Guancha. "The CCP's attempts to build a unifying ideology are challenged by highly fragmented public opinion in China's online space," wrote the Berlin think tank MERICS (Mercator Institute for China Studies) in a study looking at online discussions.[49] But in contrast to the pre-2013 years, these debates no longer reach a mass audience—and they therefore "do not represent an immediate threat to party rule," conclude the authors, though they also note that "most participants belong to the urban middle class, whose support is crucial for maintaining the stability of CCP rule."

The authors had managed to stay under the radar of the censors while conducting their survey of 1,600 forum users, yielding some surprising results. A significant 62 percent said they wanted an internationally strong China—but at the same time, despite an ongoing campaign by the CCP against "Western values," 75 percent nevertheless welcomed these values. When they were asked which values they regarded as "mainstream," they gave the top four places to freedom, democracy, equality, and individualism. It wasn't clear from the survey exactly what the respondents understood by "freedom" and "democracy." Some might have fallen for the Party's Newspeak, though the fact that socialism only came in at number seven and patriotism at fourteen suggests otherwise.

It must be noted, though, that this survey was carried out in the summer of 2016, since when the barrage of propaganda around Xi Jinping's projects—patriotism, China's superpower status, one-upmanship with the West—has increased in both firepower and intensity.

"The Chinese government has not (yet) succeeded in gaining broad-based societal acceptance, nor has it eliminated competing ideologies from the online public sphere," the study reports, before warning that, "the hardware for the ideological dominance of the Party state has been set up." Xi Jinping's ideological offensive was on the starting blocks, and the strategy would be the same as ever: the use of overwhelming power.

Another of the forums featured in the report, Gong-shiwang, where liberal intellectuals and leftists used to meet, was shut down by the censors shortly after the MERICS survey. But deleting accounts and forums is a crude tool, which makes the repressive nature of the system all too obvious. The Party now prefers to stay a few steps ahead, by working on the subjects themselves. A few years ago the *Global Times*, a nationalist tabloid sister-paper to the *People's Daily*, printed a description of the "good Chinese netizen." The ideal internet user, according to the article, should regulate himself to ensure that online "the sky is sunny and clear" and buzzes with "positive energy."

The individual citizen, to the extent that the internet has turned him into a creator, sender, and distributor of news, has also caught the attention of the censors. Every person in China, said the *People's Daily* in 2012 (in an article headlined "The internet is not a land outside the law"), is "responsible for his words and deeds." And because today, these words have the potential to travel much farther, the Party is keen to ensure that its subjects feel this extra weight of responsibility. All of a sudden, rules and laws of censorship that once only applied to the country's newspaper offices have ordinary citizens in their sights. Take the rules for administrating group chats issued in 2017, for example, which among other things prescribe "the spreading of core socialist values." They were aimed not only at the providers of chat services, but also at "those who initiate, administrate and participate in chat groups." In other words, at every individual user. The state has a direct channel into people's most private conversations. It is watching in real time, and it wants to get into your head. Watching what you write and what you do is only the first step; its real aim is for you to internalize its rules. "We need to build a 'firewall' in our brains," said the *People's Daily*; "only then can we really start talking about a foundation for national internet safety."[50]

The most widely used metaphor for China's internet controls is still the "Great Firewall," sealing off the country from the harmful influences of the outside world. And at one time, that was the aim, when China's ministry for policing launched the mother of all digital information control projects (the "Great Firewall," incidentally, would never have become so impenetrable so quickly without the lucrative—for them—assistance of Western IT and telecoms companies like Cisco, Motorola, and Nortel). Today, though, the term "Great Firewall" no longer really describes what's going on in China. It's more accurate to talk about a "Great Hive of firewalls around the individual," as David Bandurski writes in the *Hong Kong Free Press*. He sees Chinese internet surfers in the midst of a "buzzing nest of connections from which they may be insulated at will." Together, these are part of a grand illusion; people imagine "that they are part of a thriving, humming space, but all are joined to the Party's re-engineered project of guidance and managed cohesion—and all are buzzing more or less at the same frequency."[51]

This is the balancing act of clever censorship: giving people sufficient space for expression that they retain the illusion of freedom—but never so much that it allows them to grow too confident. Censorship is never perfect. There will always be loopholes, slip-ups, fine cracks in the Great Firewall, stubborn people, and free spirits who peer through the holes and cracks and glimpse the outside world. But this doesn't matter. Censorship works, and it keeps working even when it isn't perfect—as long as it's embedded within a system of lifelong mind control and manipulation. As long as the overwhelming majority of people don't even want to look at what's on the other side of the wall. As long as social controls and intimidation go hand in hand with material rewards, and people are encouraged into consumerism. As long as they have the feeling that they're enjoying more freedom than ever before.

And is that not true? The Chinese net may be flooded with sterilized information and images, but you can still drown yourself in it a hundred times over. The top ten most visited websites in China are all Chinese.* For most people, it makes no difference that the government has erased certain topics from the country's memory and banished them from public consciousness. They aren't bothered that the only news and comment you can find on government corruption, civil society, food scandals, the tensions in Tibet and Xinjiang, protests in Hong Kong, democracy in Taiwan, major natural disasters, and man-made accidents follow the government line. Many don't even notice; others just shrug and accept it.

Larger cracks in the system always become apparent when it is thrown out of balance by some unforeseen event, such as the explosion of a secret storage facility

---

* For comparison: in 2017, that figure was four out of ten in Hong Kong; in Taiwan it was five, and in South Korea three.

for hazardous chemicals in the middle of a residential area, as in Tianjin in 2015, or the spread of the coronavirus in 2020, when ultimately the horror of the events and the fear for the well-being of oneself and one's loved ones become greater than the fear of power. The veil of propaganda is lifted then for a moment for yet a new generation of Chinese, and the true nature of the system reveals itself. The system that shows itself then is one which, in case of conflict, does not hesitate for a second to place the safeguarding of its power above the welfare of its citizens. In the case of the coronavirus, it was the lack of transparency and the secrecy imposed by the bureaucracy in the first crucial weeks that allowed the spread of the disease on this scale in the first place.

In events like this, the legend of the wise dictators and the efficient meritocracy is once again exposed as a fairy tale. They reveal a clueless and irresponsible bureaucracy that under Xi Jinping has once again learned to put political obedience and ideological phrases above professionalism and actual work. Not only did the cadres in Wuhan keep their citizens in the dark about the outbreak of the disease for several weeks, they actually brought 40,000 families together for a gigantic spring festival banquet in a publicity stunt that was supposed to be a world record attempt—six weeks after the first infections and only a few days before the city of Wuhan was effectively sealed off from the outside world.

This is the deal the Communist Party has with the people: you submit to our dictatorship, and we provide you with prosperity, protection, and security in return. It is China's unwritten social contract. In view of the systemic failure during the coronavirus crisis, quite a few citizens came to the conclusion: the party had broken this contract. "It is a system that turns every natural disaster into an even greater man-made catastrophe," wrote Beijing law professor and essayist Xu Zhangrun in a widely acclaimed essay. "The coronavirus epidemic has revealed the rotten core of Chinese governance." It was a remarkable spectacle, partly because the Spring Festival holidays, and the fear of the disease, kept hundreds of millions of Chinese people in their homes, where they were glued to the screens of their mobile phones with little else to do but watch China's worst health crisis in recent decades unfold. "China's ruling elites were all of a sudden thrown into a virtual colosseum where their politician skills were mercilessly tested," wrote a Shanghai author on the blog Chublic Opinion. "And they failed spectacularly." Not only did it quickly become clear that the leaders of the city of Wuhan and Hubei Province had encouraged the spread of the disease by preferring a cover-up to an education campaign about the dangers of the virus, the local leadership elite also embarrassed themselves at ill-prepared press conferences. And while the party's propaganda machine cheered the construction of new hospitals in record time, at the same time the bureaucracy for many weeks couldn't even manage to organize enough face masks for the hospital staff.

The citizens didn't hold back their anger anymore after the death of Dr. Li Wenliang. At the end of January, the 34-year-old ophthalmologist had told former classmates in a WeChat group about a new infectious disease that reminded him of the SARS virus of 2003 and asked his colleagues to be careful in their hospitals. Thereupon the police called him in and accused him of "seriously disturbing the social order" and the spreading of harmful "rumors": "If you keep being stubborn, fail to repent, and continue illegal activities, you will be brought to justice," the police warned the young doctor. Until then, Li Wenliang himself had been something of a model citizen of the system, certainly not a dissident, and if you look closely, even the term "whistleblower" was a bit of a stretch for his actions: he had not gone to the press, had not posted anything publicly; the only thing he had done was to warn friends and acquaintances—doctors like himself—in a semi-private forum of a health risk. Even that can count as a criminal act in China today.

His death turned Li Wenliang from hero to martyr. Many internet users reposted this sentence from an interview with Li shortly before his death: "I think a healthy society should not have only one voice." Immediately, hashtags like "#We-want-freedom-speech" or "#We-demand-freedom-speech" were shared millions of times in China's social media. It was probably the biggest outburst of popular anger since the 1989 Tiananmen Square democracy movement—and it took place entirely in virtual space: nobody took to the streets, if only for fear of contagion.

For the party it was horror: the loss of control caught them cold. But it reacted quickly with its tried and tested instruments: censorship, propaganda, repression. There had been citizen journalists, video bloggers such as the lawyer Chen Qiushi and the businessman Fang Bin, who had posted pictures of chaos from Wuhan's hopelessly overcrowded and overstretched hospitals on social media. Both Chen and Fang disappeared in early February in the hands of the security apparatus. The freedom-of-speech discussions were quickly blocked and erased. Censorship found itself in a dilemma, however: by now the central government had stepped in, and propaganda now demanded the education of the public about the epidemic, so censors could hardly erase all debates about the coronavirus. Instead, one instruction by the censorship authority to the country's editorial offices read that "the temperature of the debate be safely controlled."

At the same time, propaganda started a counteroffensive: the party had declared a "people's war" against the virus, in which everyone would now have to stand together and act as one (one more reason to not allow dissenting voices). State TV, newspapers, and websites were now full of pictures of the new hospitals pounding out of the ground, with stories of heroic self-sacrifice among doctors and nurses. In general, the Xinhua news agency advised: "A positive, confident attitude is the best immunity against the epidemic."

It took Xi Jinping a remarkable two months after the first cases of infection to appear publicly on the CCTV evening news as the top crisis manager for the first time. It immediately became clear that for him, the freedom of speech demanded by millions of citizens online was by no means part of his solution for the crisis. He read out a list of the measures that had to be implemented immediately in the fight against the epidemic, and already point 3 on that list was a demand to his cadres that they had to "comprehensively strengthen social control" and "maintain social stability." Point 4 read: "We must strengthen the guidance of public opinion." The media, he said, should now focus on "the great love of the party," as well as on "the touching stories of front-line medical personnel, grass-roots cadres, public security police, and community workers."

A few days later, the authorities identified ten categories of crimes related to the epidemic, which were to be severely punished. These categories included not only price gouging, the hoarding of face masks, and illegal trade in wild animals but also "malicious fabrication of epidemic information, causing social panic, stirring up public sentiment, or disrupting social order, especially maliciously attacking the party and government, taking the opportunity to incite subversion of state power, or overthrow of the socialist system."

For the younger generation, who had not experienced Tiananmen 1989 or SARS 2003, the epidemic was a shock: it was the first time the system had revealed itself to them in such a gloomy light. When Deputy Prime Minister Sun Chunlan visited Wuhan in early March to prepare for the visit of Party leader Xi Jinping, the citizens were still so livid that it tore through the image of the well-staged propaganda tour. In a video clip circulating on the net, afterward one could watch Sun's inspection of a housing estate that had been prepared as a perfect Wuhan middle-class idyll—well, almost perfect. The citizens of the city were locked up in their apartments and were not meant to be part of this production, but they managed to make their anger heard. Just as Sun, accompanied by men in uniform and her delegation, is walking down the empty and peaceful inner yard of the housing estate, a choir of screams suddenly tears through the silence: "*Jiade! Jiade! Dou shi jiade!*" shout the locked-in citizens from their balconies. ("Fake, fake, everything's fake!")

After the disaster of the first few weeks, the CCP had indeed pulled itself together in an astonishing show of strength. Within a very short period of time, it quarantined 60 million inhabitants of Hubei Province and banned hundreds of millions of Chinese from traveling and moving about. The breathtaking efforts of the Chinese health and security authorities were accompanied by an almost similarly astonishing propaganda onslaught. Seldom before has left become right and top become bottom so quickly. The Party had finally declared "total war" (*Xinhua*) on the virus after weeks of cover-up and now it declared total victory. Instead of reviewing the system failure of the first weeks and learning from it, the Party apparatus quickly declared the CCP and its system the triumphant victor of the crisis: Its "institutional strength" had been key to China beating the virus, wrote Xinhua: "Under the leadership of the Communist Party of China, people from all walks of life have joined hands in fighting the epidemic with wisdom, action, and morale. All these are vivid manifestations of China's system advantage."[52]

One of the few voices in those days who still dared to counter the astonishing new propaganda narrative with a reminder of the very real party failure only a couple of weeks before was that of 69-year-old Ren Zhiqiang. As an entrepreneur, party member, and descendant of a powerful family, Ren himself used to be part of the CCP elite before Xi Jinping. The essay that he wrote at the end of February and distributed to friends was an attack on Xi Jinping and the Party leadership, whom he accused of wanting "to cover up their own scandal."[53] He describes how they propagate fairy tales of the CCP's glorious struggle while at the same time not being capable of a single word of self-criticism. There was no investigation into what really happened in Wuhan, no naming of those responsible: "It's just as if those in power don't want to accept any responsibility, and are refusing society's wish to know who is responsible." In reality, Ren wrote, the country was governed by an "emperor without clothes" under whose rule the system had become "seriously ill" and as a result was now creating "collective harm." Ren's essay was widely circulated on the net, shortly after which the 69-year-old disappeared. In early April, the CCP declared that Ren was under investigation for "serious violation of party discipline and the law."

On March 7, the newly appointed Wuhan Party secretary Wang Zhonglin declared that what the people of Wuhan now needed was a "gratitude education." After all, he said, the residents of the city, should now "thank the General Secretary, and thank the Chinese Communist Party." This was still a risky step at that time: users on Weibo and WeChat mocked and attacked Wang for his gratitude project in comments on social media that were quickly deleted. A couple of weeks later, the Communist Party declared the late whistleblower and hero of Wuhan's citizens, Dr. Li Wenliang, one of their own. The Party, whose officials had threatened and silenced Li only a couple of weeks before, now praised him as an exemplary Communist and officially raised him to the status of martyrdom at the beginning of April. "'Martyr,'" explained the *Global Times*, "is the highest honorable title awarded by the Communist Party of China and the country to its citizens who died bravely in service to the country, society, and the people."

Will the shock last? The anger that feeds on all the feelings of fear, powerlessness, and insecurity that citizens have experienced and for which this system leaves them no other outlet than to scream from their balcony? There are signs that the crisis was a political awakening especially for some young Chinese, such as those who have set up online archives such as the encrypted channel in the app Telegram they called "Cyber Graveyard." There thousands of users collect articles and comments that have been deleted and blocked by the censorship. These are archives from which an uncensored history of the pandemic in China can be written in the future. The ever-recurring amnesia decreed by the party is being subverted.

At the same time, the propaganda quite successfully feeds into a new nationalist zeal. It does so by constant reference to the chaotic virus crisis management in the US and Italy, but also with crude disinformation. For example, according to one piece of news the Chinese were being fed even by some of their Foreign Ministry officials, the virus in reality did not come from Wuhan but was created in a US bioweapon laboratory. For the time being, this propaganda is likely to catch on with the majority of citizens. How lasting this propaganda's success will be will depend on many factors, especially how much longer the strong growth of China's economy will allow the Party to buy the goodwill and good conduct of its citizens. The party apparatus is still strong, and its instruments of thought control have been tried and tested many times. It has gone through many a similar crisis in the past, only to have its propaganda declare full scale victory once the immediate threat subsided.

After a few weeks in Hong Kong during that summer of umbrellas, when the people occupied the center of their city, intoxicated by their new-found strength and solidarity, I flew back to Beijing. I went straight from the airport to a party held by a painter friend in the north of the city. A few artists had turned old chicken coops into studios there; they were barbecuing skewers of lamb and there was beer and wine. Hong Kong dominated the conversation, and the question of why almost all Chinese people were blithely parroting their government's propaganda, declaring that the students in Hong Kong were being ungrateful to their fatherland, and dancing to the tune of shadowy foreign powers.

Most of those present were in their fifties and had lived through the Cultural Revolution—an experience that turned many of their generation into lifelong skeptics. Just before the party, one of the guests, a painter from the Songzhuang artists' quarter, had received the news that some of his friends had been arrested that morning for planning to attend a poetry reading in support of the Hong Kong students. He spoke of the nationalistic and militaristic education system that the Party had rolled out across the country in the wake of the Tiananmen Square massacre in 1989, which was now being developed further under Xi Jinping. It was having the desired effect, the painter said. "People born in the 1980s and afterward are hopelessly lost. The brainwashing starts in nursery school. It was different for us. They called *us* a lost generation because schools and colleges were closed back then, and many of us were denied an education. But in reality we were probably the lucky ones. We fell through the cracks. The brainwashing didn't get us. Mao was dead, and everyone was desperate for China to open up, for reform, freedom."

David, a 28-year-old native of Beijing who teaches English at an elite grammar school there, says of his 17- and 18-year-old students: "They're incredibly tech-savvy. And they use tunnel software and other technologies to get past the censors much more than we do. They also use it to access banned sites like Facebook and YouTube—but they do so purely for entertainment, to follow celebrities." David tells me that he sometimes gives them English books to read in his lessons, especially works of history, and sometimes articles from the *New York Times*. "But they look at me helplessly. Their thinking isn't joined up anymore, they don't have any background knowledge." This is a generation who with just a few more clicks could access all the information in the world. But they don't do it. They don't want to.

"My students say they haven't got time. They're distracted by a thousand other things," says David. "And although I'm only ten years older than them, they don't understand me. They live in a completely different world. They've been perfectly manipulated by their education and the Party's propaganda: my students devote their lives to consumerism and ignore everything else. They ignore reality; it's been made easy for them."

As the cultural critic Neil Postman, author of *Amusing Ourselves to Death*, once wrote: "What Orwell feared were those who would ban books. What Huxley feared was that there would be no reason to ban a book, for there would be no one who wanted to read one."[54] Weibo still exists. It's bigger than ever: with more than 490 million active users a month in 2019, it is even bigger than Twitter with its 320 million—and it's still growing. But people aren't reading, watching, and commenting on the same things they were a few years ago. The most-shared Weibo post of all time dates from September 2014: "I am 15 today, and so many of you are on my side. Thank you for always being with me these last few years." It may well be the most-read social media post in the world, with more than 100 million shares.

Its author was one of the TFBoys, Wang Junkai. The TFBoys, a manufactured boy-band, are probably the most popular group among China's young people over recent years. Three lads with an image somewhere between chorister, K-pop, and communist Young Pioneer, with any hint of rebellious youth carefully blow-dried out of them. They sing about working hard at school, being a team player and serving the motherland. In one video, they dress up as Young Pioneers and sing: "We are the heirs to communism." They have 30 million followers on Weibo; add in the songs, the TV series, and the films, plus merchandise, and these coiffured and primped Marxist pop-princelings are netting the equivalent of several million dollars every month. This is what China's social media looks like today.

Every year there is one day that pushes China's internet to the limits of its capacity; a day when people all over the country sit in front of their screens from midnight until midnight, clicking for all they're worth. "Singles Day" is a marketing coup by the online marketplace Alibaba. In the space of just a few years—with the help of enticing discounts and bargains—those 24 hours have been turned into one of the most important holidays in China. It is the most unbridled celebration of consumerism anywhere in the world.

"Singles Day"—November 11—was launched as a joke by frustrated girlfriend-less Nanjing students, but Alibaba has converted it into a day-long orgy of online shopping, which has long since overtaken its precursor and role model—the USA's Black Friday. On November 11, 2019, Alibaba turned over the equivalent of more than $38 billion; Amazon's turnover for Black Friday and Cyber Monday combined was less than half that amount, at almost $14 billion. The singer Taylor Swift and Hollywood stars like Nicole Kidman and Daniel Craig have all appeared at the annual Alibaba galas in Hangzhou which kick off the day. In the aftermath, China's postal service has to deliver a billion parcels. The Beijing branch of Greenpeace may have called Singles Day a "catastrophe for the environment," but it hasn't stopped the nation descending into its annual 24-hour frenzy. Global communism is dead, and—thanks to the net—global consumerism has opened its new HQ in China.

How far China has come from the days—not so long ago—when food was rationed and you needed special coupons to buy a bicycle or a typewriter. Today consumption is a pleasure that the Party not only permits but encourages its subjects to exercise without restraint. Capitalism has shown them the way: consumption brings growth, consumption is a sedative. This logic is applied even more deliberately and shamelessly in China than elsewhere. "The world's stable now," says the Controller in Aldous Huxley's *Brave New World*. "People are happy; they get what they want, and they never want what they can't get." The perfect totalitarian state, Huxley wrote in a 1946 preface to his book, is one in which the all-powerful caste of rulers "control a population of slaves who do not have to be coerced because they love their servitude."

Huxley felt that the earliest advocates of universal literacy and a free press had lacked imagination, envisaging "only two possibilities: the propaganda might be true, or it might be false. They did not foresee what . . . has happened, above all in our Western capitalist democracies—the development of a vast mass communications industry, concerned . . . neither with the true nor the false, but with the unreal, the more or less totally irrelevant. In a word, they failed to take into account man's almost infinite appetite for distractions."[55]

As Huxley wrote these words, he was witnessing the

first excesses of modern capitalism. Donald Trump's America and Xi Jinping's China are a testament to his great prophetic power.

At the start of 2018, Stanford University published a fascinating study that supported the conclusions reached by David, the teacher quoted above.[56] The authors David Yang from Stanford and Chen Yuyu from Peking University studied the internet usage of more than 1,800 Beijing students between 2015 and 2017. Before the start of the project, 80 percent of the students had never tried to beat the Great Firewall—some of them partly for reasons of cost. Virtual Private Networks (VPNs) allow you to get past censorship and restrictions on the internet using "VPN tunnels," but many VPN providers charge for their services. Across the whole country, the authors estimate, only between 1 and 8 percent of internet users regularly use this software. (Bear in mind that this research took place before the Party's vigorous campaign against VPN services in spring 2018.) In order to address this issue, the researchers gave the participants cost-free access to tunnel software for 18 months.

Even after being reminded six times, only 53 percent of the students activated the software. The authors of the study were well aware that fear could be a factor in the choices the students made. So they deliberately muddied the waters by offering free access to a permitted movie-download site alongside censored domains; this helped to cover up the true focus of their study. One of their interests was to see how many students would take the opportunity to read foreign news websites blocked in China. Would they, for example, take a look at the Chinese-language edition of the *New York Times*? Those who did so made up less than 5 percent of those students who had activated their VPN tunnels.

This result had nothing to do with inadequate foreign-language skills: Taiwan, Hong Kong, and the USA have thousands of Chinese-language websites offering information outside the Beijing propaganda matrix. But fewer than one in forty of these young people felt the need to access such information. And this was a group of students at one of the most famous and formerly most liberal universities in China. The best-educated young people in the country; China's future elite. Censorship doesn't just work because the regime makes it difficult to access free information: "Rather, it fosters an environment in which citizens do not demand such information in the first place."[57]

How could the researchers be sure, though, whether the subjects of their study were incurious, as a result of the sterile, censored world in which they'd been raised, or afraid of what would happen if they dared to look beyond it? The second phase of the study was designed to address this point. For one group, the researchers devised quiz questions that they could only answer by reading the *New York Times* and other sites, rewarding each correct answer with a small cash sum of around $2.50 (this was deemed unlikely to sway students genuinely frightened of the secret police).

So if these students responded by spending time on uncensored Western sites, then something other than cash must be motivating them. And they did indeed spend time. By the end of the study, the percentage of waking hours the group spent on websites like the *New York Times* was nine times greater than it had been at the start. After a while the students began to search for information the researchers hadn't requested; and they started making regular use of Wikipedia, which is blocked in China.

As soon as the students were confronted with the fact that something exciting and valuable was hiding on the other side of the wall, their interest grew rapidly. Among the students who were now regularly consuming information that was banned in China, the researchers saw "broad, substantial, and persistent changes to students' knowledge, beliefs, attitudes, and intended behaviors."[58] Their trust in China's government and its institutions fell significantly, and they displayed a growing skepticism and mistrust. They became more pessimistic in their assessment of China's economic development. Many expressed a belief that China's political and economic systems needed a fundamental overhaul.

So the study held good news and bad news for the Party. The bad news was that it really does need to fear the free flow of information. The good: as long as it continues to tailor its system of mind-and-information control to developing technologies and make use of them for its own ends, it has nothing to fear for the time being. Its censorship doesn't have to be perfect; most citizens will not go looking for the loopholes.

The researchers from Stanford and Beijing concluded that the demand for free information "is not low by nature, and that it also is probably not the fear of being punished by the authorities that leads to students not demanding sensitive information." It was their "failure to value uncensored information" that dampened their curiosity. People don't feel like they're missing out. They're like the frog in the Chinese fable, who thinks his well is the whole world and the small circle of sky that he can see from the bottom is the entire firmament. And when the turtle happens upon the well one day and tells him of the boundless sea, the frog looks at him with incomprehension: he can hop about on the edge of the well, rest in the cracks, take a mud-bath at the bottom—doesn't the frog already have every happiness the world can offer? What does he want with this funny sea-thing the turtle is talking about?

A good proportion of the people existing under such a system remain immune to information and arguments from outside, even when they do get through. Just as a mayfly cannot discuss ice and a narrow-minded scholar cannot discuss life, said Zhuangzi, the philosophical forefather of Taoism, a well-frog cannot talk about the sea.[59] Perfect mind control has been achieved when the frog carries his well around with him. He can walk out into the world and stand beneath the vast skies, and still see only the small circle in which he grew up.

The Clean Sheet

Why the People Have to Forget

"The past is never dead. It's not even past."
William Faulkner

The Party has good reason to celebrate. June 4, 2019, is the thirtieth anniversary of the massacre on Tiananmen Square. This wasn't just the end of the democratic movement, it was also the end of a great festival, a Happening at which millions of citizens got high on their new-found freedom and their dream of a better China. "I don't know what we want," one of the students in the square had shouted exuberantly. "All I know is, we want more of it." What they got, on the night of June 3–4, 1989, were bullets and bayonets. Tanks rolled in and made the night shudder. Students, workers, and bystanders were crushed, shot, and stabbed in their hundreds or thousands; we still don't know the exact figure. In hindsight, and from the Party's point of view, it was a success. A greater success than anyone could have imagined at the time.

"The Chinese had to learn to forget, in order to survive," said one man who decided to remember. The painter Zhang Xiaogang is a survivor. The period leading up to 1989 was the freest time the People's Republic of China had ever known. The people had escaped the horrors of the Cultural Revolution, and Deng Xiaoping, the new strongman, threw the windows to the world wide open. To start with, he didn't care whether "a few flies" came in with the fresh air, to use his own phrase. Zhang Xiaogang remembers that time well. "There was suddenly so much hope, so much illusion, so much beauty. It was the most poetic time. A time when the people queued outside the Xinhua bookshops to buy a novel."

1989 was a remarkable year. Zhang was working on a painting of a red woman on the banks of the classical river Lethe. In the Greek myths, the waters of Lethe give anyone who drinks them the gift of complete forgetfulness. In January of that year, the now legendary "China/Avant-Garde" exhibition opened in Beijing's national museum, bringing China's new art out from the underground for the first time. The exhibition opened its doors at 9 a.m., and by 3 p.m. they were being closed again by the police: nervous culture-department bureaucrats had seen something too unfamiliar and wild for their liking. A new generation of artists was speaking a language they simply didn't understand. And yet: "We were there now," says Zhang. "We thought China would inevitably become more open and freer." For just a few more precious months, they would hang onto their beliefs, their idealism, their innocence.

Then came that night, June 4, when the tanks rolled in, and the China we know today was born. All the blood and terror that followed saved the CCP at a time when socialist regimes around the world were crumbling into dust. With the defeat of the "counter-revolutionary riots," Deng Xiaoping secured his rule and that of many senior officials who had been condemned as corrupt and nepotistic by the demonstrators—and today are the heads of family clans in possession of wealth unimaginable in the late 1980s. The massacre bought the regime valuable decades. It gave Deng back power over truth and memory. And in the end, it won him admiration from politicians and businesspeople in the West. As Bertolt Brecht once asked: "Would it not be simpler/ for the government/ to dissolve the people/ and elect another?" China's government didn't bother with elections; it shaped another people for itself.

The astrophysicist Fang Lizhi was one of the heroes of the 1989 movement. After the massacre he sought comfort in his certainty that, in light of the monstrosity of those days, in light of the millions of witnesses, this one time at least, "the technique of forgetting history was doomed to fail."[60] Few people could have guessed just how wrong Fang would prove to be. The CCP simply pressed "delete" and reformatted the Chinese people. The Tiananmen Square massacre shook the world—but in China, it has been forgotten. The year 1989? "Without doubt, quite an ordinary year," according to a poem by the writer Yang Lian.

Guns and pens work together. The state's soldiers murder the protesters; the state's writers murder the truth. "Who controls the past controls the future. Who controls the present controls the past," wrote George Orwell. Mao Zedong once wished his people were a blank white sheet of paper, upon which "the freshest and most beautiful characters could be written." The autocrat needs the collective amnesia of the people. The Party's propaganda never ceases its task of reordering and reconfiguration. After June 4, 1989, the country was frozen in shock. When the people finally shook themselves out of their stupor and started bowing and scraping again, the propaganda machine began to artfully edit out the horror of those days. Over the years the "counter-revolutionary riots" became known first as the "riots," then the "political storm," and finally just "the incident."

In the end, even the "incident" dissolved into silence. As if an old photograph had faded until only meaningless silhouettes remained. When Beijing police officers today warn foreign correspondents against writing about the massacre, they couch their warnings in vague terms, referring to "sensitive issues in sensitive times." They aren't even specific enough to say: *that year* or *that day*. In fact, every year around June 4 the very terms "that year" and "that day" reappear in the internet censors' index of banned words. On Weibo, blocked expressions also include: "When spring becomes summer" and "35 May," terms thought up by inventive users as proxies for "June 4." Wikipedia is banned in China; its Chinese counterpart is called Baidu Baike, and claims to be "an open and free online encyclopaedia." You will find entries on Baidu Baike for the years 1988 and 1990—but 1989 doesn't exist. An entire year has been erased from history.

The mind-blowing thing about this is not the efforts of the censors, but the fact that they have worked. It can be done. Even in an era when, thanks to the internet, China's doors were more open to the world than they had ever been, it was still possible to brainwash an entire nation. Thousands of young, intelligent, open-minded Chinese people, who have grown up with smartphones and social media, go abroad at twenty and hear for the first time the news of what happened in their homeland in 1989. Some are shocked; many others refuse to believe it even then. They close themselves off inside the cocoon the Party has spun for them, and refuse to let the information get to them. "Why won't you Western journalists stop lying?" a Chinese student in Germany once wrote to me. "You just can't bear the fact that China is growing strong."

The civil rights activist Hu Jia tells the story of January 17, 2005, the day Zhao Ziyang died. Zhao was the liberal head of the Communist Party from 1987 to 1989, and prime minister* prior to that. It was Zhao who tried to make some concessions to the students on Tiananmen Square, before being toppled by the hard-

---

* The Premier of the State Council of the People's Republic of China, sometimes also referred to informally as the "prime minister," is the leader of the State Council of China (constitutionally synonymous with the "Central People's Government" since 1954), who is the head of government and holds the highest rank (Level 1) in the Civil Service.

liners around the *eminence grise* Deng Xiaoping. Hu Jia knew his family and went to offer his condolences. When he got home, his wife Zeng Jingyang asked where he had been. He explained. She gave him a quizzical look: "At whose house? Zhao Ziyang? Who's that?"

"That was a bit of a shock for me," says Hu Jia. "She was born in 1983, and she's a clever, critical woman who studied at the People's University, one of the elite universities in Beijing. And she had never in her life heard of Zhao Ziyang, who was—at least nominally—the most powerful man in the country for several years. At that moment I understood the power the Party has over our brains."

When the American journalist Louisa Lim was doing research for her book *The People's Republic of Amnesia*, she showed a hundred Beijing students the famous photo of "Tank Man": the man in the black trousers and white shirt, standing in the way of the approaching convoy of tanks with nothing but a plastic bag in his hand, and bringing them to a halt just an arm's length away from his slender frame. It is one of the iconic images of the 20th century. But only 15 of the 100 students recognized—and to their extreme dismay—that the image was of China. Beijing. The Street of Eternal Peace that leads to Tiananmen Square. The other 85? They shrugged and guessed Kosovo or South Korea.[61]

The business of mental eradication—forgetting—is made easier when every few years, every few weeks even, the physical world around you is also pulverized, vanishes without trace and is completely reconfigured. In the past two decades the whole of China has been torn down and rebuilt, in some places several times over. This has brought a Western style of living to China's cities, with beautiful new apartments in tower blocks that have sprung up overnight, and modern, faceless towns. At the same time it has erased everything that was familiar to people.

"You just had to turn your back and the old familiar street would be gone," says Zhang Xiaogang, the painter, talking about the 1990s. "I was away from Kunming for a semester. Then I came back—and my hometown didn't exist anymore. Destroyed, laid waste overnight." Wherever you went at that time, China's cities had been reduced to rubble. They were waiting to rise again as thousands of clones. "I think it's the first time this has happened, at such a speed, on such a scale, anywhere in the world," says Zhang. "Everything in this country is being radically changed day after day. In your city, but also in your life. The speed at which things are happening here is abnormal. It goes far beyond what the human soul can normally bear."

The wiping of memories through cityscapes and architecture is just one expression of a collective forgetting that has become an instinctive survival technique for the Chinese. "The old buildings were probably the last things to be rubbed out," says Wang Shu, China's most famous architect, who has won the coveted Pritzker Prize. "That explains why people stopped objecting to them being demolished and completely rebuilt. Everything else, our old life, our tradition and culture, had already been shattered, particularly during the Cultural Revolution. The old houses were just a final, hollow remnant."

Today, China has more than 100 cities bigger than Chicago. Many of them have grown to ten or more times their original size over the past decade, and all have completely eradicated their history in the process. "It's as if someone dropped 120 nuclear bombs on China," says Wang Shu. Of the buildings he designed in the 1990s, not a single one is still standing. They have all been torn down.

Sometimes, among China's desperate, tumbling population, you can sense a lost-ness; a great insecurity. Both Zhang Xiaogang the painter and Wang Shu the architect had decided early on to try to keep their balance by stubbornly remembering. This takes a lot more effort in China than it does elsewhere. It also makes you an outsider.

The two men gained recognition within China only once their successes abroad had made them famous. Zhang became known for his family portraits, which were inspired by studio photos from the Cultural Revolution: people wearing cotton tunics, as stipulated by Mao, and serious, unmoving expressions. Almost everyone who had an interest in China or modern art was captivated by his portraits, and the hypnotic eyes of these dream-like faces. They are serious faces, calm as a silent lake, but their mute resignation hides a storm of emotions; the pain and desires of an individual human being swirl below the surface. All the pictures are slightly out of focus; the only living detail is the shimmering black eyes.

Zhang Xiaogang gave the title "Amnesia and Memory" to a series of pictures begun at the start of the new millennium. The painter sees the sheer force of the change that has been unleashed in China as something that you have to resist to have any hope of holding onto the past. But in doing so you are up against the might of the Communist Party, which views memory as a dangerous and subversive power. If you make people homeless and rootless, it calculates, they will run, naked and shivering, into the protective arms of the Mother Party.

The inspiration for Zhang's work comes from his

childhood during the Cultural Revolution. From the age of five, he was trapped inside the four bare walls of the Chengdu apartment allocated to his family by the work unit. Just like the little boy in one of his paintings, who rides his bicycle around a table. No light shines through the windows in these pictures. "All the windows in our building had been bricked up," he remembers. They stayed that way for two years, because of the Red Guards who occasionally fired shots at the building. There was no electricity either; just oil lamps.

"It was dark. Like a bunker. My eyes were ruined." Was it a terrible time? Yes and no. The eight-year-old Zhang and his friends thought it was cool. "We were a big gang of friends. And there were no adults there to keep an eye on us. We roamed around just as we pleased." The Cultural Revolution was one of humanity's darkest hours. But for some of the children—you hear this quite often—it was tremendous fun. The adults were out at "struggle sessions" from morning until night, denouncing and humiliating each other. The Red Guards paraded around triumphantly with the bodies of "counter-revolutionary enemies" who had been tortured to death. Meanwhile the young children raced up and down the stairs, finding themselves in a fairy-tale realm of anarchy where all the adults were far, far away. They may have been freer and less constrained than any other generation of Chinese children. No authorities, no school, no homework, for years. Zhang saw a lot of dead bodies in those days, and heard a lot of shooting. "It was terrible for a little boy," he says. "And exciting."

From 1966 to 1976, the Cultural Revolution laid waste to the nation's souls. "And today," says Zhang, "it has been forgotten." This was particularly obvious in 2016, exactly fifty years since Mao fired the starting pistol. There was not a single commemoration all that year, not a single memorial event. No debates, no self-criticism. A blanket of silence lay over the country. The Party knows that it might be under threat if people remember too much. So it rewrites the past into a fantastical historical novel that serves its present needs.

After Mao's death, the Communist Party grandees condemned the Cultural Revolution as "ten years of chaos"—but they quickly added that Mao had done the Motherland a great service, and his errors were trivial in comparison. The Party drew up a balance sheet and concluded that Mao had been "70 percent good and 30 percent bad." The 30 percent part would include the 40 million people who died of starvation during the "Great Leap Forward" from 1958 to 1961—that devastating campaign in which Mao ordered every last peasant in every last village to melt down their tools and pots to provide steel. Mao wanted to "catch up with England and overtake America" by turning China into an industrial nation overnight. Eventually, China had no spades or shovels left, no ploughshares and no woks, and the result was one of the greatest famines in history.

Not long after taking office, Party leader Xi Jinping declared that any criticism of Mao Zedong was once more taboo. What the Germans call *Vergangenheitsbewältigung*—coming to terms with the (Nazi) past—the CCP calls "historical nihilism," and those who practice it can be prosecuted. In April 2018 the National People's Congress passed a law threatening punishment to anyone "insulting or slandering the [communist] heroes and martyrs"—meaning anyone who dared to question the historical accuracy of the heroic stories from the Party's glorious past. On the day I went to visit the painter Zhang Xiaogang in his studio, some workmen were replacing the famous portrait of Mao at the gate to Tiananmen Square: the Great Chairman was to appear untarnished and in fresh colors for the national holiday.

"I always thought history and memory would triumph over temporary aberrations and return to their rightful place," the author Yan Lianke wrote in 2013.[62] "It now appears the opposite is true. In today's China, amnesia trumps memory. Lies are surpassing the truth." Fear is one factor. The silence of those who were there, the mute acquiescence with those in power: in China these are simple survival techniques. You fear for your children and hide the truth from them, because—spoken unthinkingly in front of classmates or teachers—it might put them in danger. I once asked Cui Jian, the rock musician who had played to the students in Tiananmen Square in 1989, and whose song "Nothing to My Name" (*Yi wu suo you*) became their anthem, what he was most concerned about in China. "*Zhuang-sha*," he replied like a shot: the pretense of ignorance that becomes second nature to people living under dictatorships.

Yan Lianke decries the silence and the timidity of his fellow writers and the caste of intellectuals as a whole when faced with this state-led amnesia. "The opened window we have now is more a small mercy bestowed by the powerful at a permissive moment, than a victory won by intellectuals because of our persistent quest for openness," he writes. "Someone who has spent years in a dark cell is bound to be grateful if a window in his cell is unshuttered and some light is allowed in. Would he dare to ask for the prison gate to be opened for him?" Yan Lianke keeps returning to this subject in his writing. He did so in an e-lecture for Hong Kong students at the height of the coronavirus crisis, when back in Beijing the Party had once again begun to ask the citizens to praise its wisdom and greatness. He often thinks, writes Yan Lianke, that memory is more important to people than food, clothing, or breathing: "Bodies have not turned cold and people are still mourning. Yet, triumphal songs are ready to be sung." Yan implored the students not to let anyone force them to forget. "Forgetful people are, in essence, dirt in the fields and on the roads. Grooves on the sole of a shoe can step on them in whichever way they please. Forgetful people are, in essence, woodblocks and planks that have cut ties with the tree that gave them life. Saws and axes are in full control of what they become in the future."[63]

But there is more than just fear at work here. After 1989, something new was added to the mix. The government of Deng Xiaoping offered its people (the city-dwellers among them, at any rate) a deal: make money, get rich—but keep your mouths shut. And the people accepted that deal. *Wang qian kan* was the slogan of the moment: "Look ahead!" Just don't turn around. Helpfully, the sounds *wang qian kan* can also be written with different characters that mean "Follow the money!" Since then, the Party has been pursuing an unprecedented course of economic opening-up, while refusing to loosen its political stranglehold. It still worships the shrines of communism as if they contained holy relics, although everyone knows that the thing they're required to venerate has been dead as a doornail for years. The leaders have hit upon nationalism as the new opium of the people, a poisonous little plant that they have been conscientiously feeding and watering, and which has now produced a host of wild blossoms.

The past may be forgotten, but it is not past. The way China's society functions today cannot be separated from what happened in the Cultural Revolution. The Cultural Revolution, says Zhang Xiaogang, isn't just an historical fact, it's more a "mental state." Today's China may be glittering and booming, but you shouldn't let that deceive you: "Much in the psyche of Chinese people today goes back to that time." A trauma of this kind, as we know from studies of war, is passed down the generations. If everyone in China today mistrusts everyone else; if everyone automatically assumes that everyone else is out to cheat and trick them, that attitude is rooted in a time when husbands betrayed their wives and children got their parents sent to labor camps, or even to the scaffold. The young people of that time, both perpetrators and victims, are now in power. They're leading the Party, the state, the big Chinese businesses.

The way China looks today is also impossible to

separate from what happened in 1989. The glittering, powerful, ambitious China—and the rottenness at its core. The boom and the growing prosperity that Chinese cities have enjoyed for many years has turned the urban middle classes into the Party's most loyal accomplices. And because ideals have been banned in the country since 1989, they have thrown themselves into uninhibited materialism. The families of the CCP's leaders have enriched themselves shamelessly in the process, foremost among them the clan of Li Peng, who as prime minister declared martial law in 1989. This is a society in which young people's idealism and willingness to make sacrifices have now slid into a general cynicism. The moral crisis ravaging China's society, people's loss of all trust in each other and in the state apparatus, the rampant corruption, the ecological devastation—all these things can to some extent be traced back to the course set in those fateful moments.

When you have been ordered to forget what happened on the night of June 4, then remembering is a crime. Year after year, in the early days of June the Weibo censors' index of banned words includes *daonian*, "mourn." Mourning is forbidden, in word and deed. Who exactly is the Party afraid of? Clearly the handful of stubborn people who cling to their memories. Each year the performance is repeated: in the weeks leading up to June 4 the authorities silence dozens of these stubborn people or throw them in prison. The drawing Zhang Xiaogang posted on his Weibo account on June 4, 2012, of a face with hands clapped to its mouth in horror, was deleted immediately. In spring every year, foreign journalists and diplomats start finding it almost impossible to contact the "Mothers of Tiananmen Square," a group of female democracy activists who refuse to stop commemorating the violent deaths of their sons and daughters. And then there is Chen Guang, one of the soldiers from that time who later became a painter: he was arrested just before the 25th anniversary of the massacre, after a performance for friends during which he painted the numbers 1989 to 2014 on brightly colored walls, and then whitewashed over them. Activist Chen Bing and his brother took part in the 1989 protests, and, in 2016, together with friends, produced a homemade "Tiananmen memorial" schnapps priced at 89.64 yuan; the label on the bottle showed a photo of the "Tank Man." A few months later, in April 2019, a court in Chengdu sent him to prison for three and a half years.

The massacre of 1989 has not driven everyone into silence. For many in the small band of people who might be called China's conscience today—lawyers, civil rights activists, intellectuals, authors—that night was the trigger for their activism. But overall, the propaganda campaign has been a resounding success. "Nobody below the age of fifty has any real idea about the things that happened under Mao," says the journalist and author Yang Jisheng. "They think he was great and good, because that's what television and books tell them."

Yang Jisheng used to be a journalist for the state news agency Xinhua, and he used his status to gain access to archives all over the country. After many years of research, he wrote the book *Tombstone*, in which he revealed previously unknown details of the "Great Leap Forward."[64] Yang presents evidence that cannibalism took place in many parts of the country, and estimates the number of deaths at 36 million; other historians such as Frank Dikötter put the figure at up to 45 million.[65]

Sometimes selective acts of remembering are allowed, as with the Cultural Revolution: years later, isolated books and films were permitted to describe, on the level of personal experience, the chaos and cruelty of those years—though at all times, the question of *why* remained taboo. How could this happen? Who was responsible? What role did the Party play? A Party which had brought forth Mao in the first place, and whose apparatus enabled his brutal power game to take place, at the cost of an entire nation.

The Cultural Revolution flowed into a cruel campaign to eradicate the "Four Olds": old thinking, old culture, old habits, and old customs. In the end, it managed to destroy everything the old Chinese culture had accomplished—preserving for the new age only the old evil of Chinese despotism. Ironically, today's Party leadership exploits the memory of the deadly turmoil of those times (memories that it has at other times been at pains to suppress and distort): "Chaos must never reign again!" cries the propaganda machine, knowing that in this at least it can count on the agreement of its traumatized citizens. And in the next breath: "Stand united behind the leader and do not doubt him!" Its mantra is "stability." Repression of every kind comes under the heading of *weiwen*, "maintaining stability."

Unlike the Cultural Revolution, the "Great Leap Forward" has been completely forgotten—despite all the human lives it cost. If the Party ever mentions that period, it speaks coyly of the "difficult years." The Party prescribes historical amnesia to the people over and over again; otherwise it would have to face up to its crimes.

The hatred of Japan, and the endless repetition of

Japan's crimes against the Chinese people, is one of the central themes of its nationalistic propaganda. What might happen if the nation realized that the CCP under Mao Zedong, according to US sinologist Perry Link's estimate, probably "killed about eight to 10 times more Chinese people than the Japanese did?"[66] "One might say that the Japanese were crueler, with their head-lopping contests and live burials during the Nanjing massacre in 1937," Link writes, "but who can say whether that was crueler than the gouging of eyes during the Cultural Revolution or, in Guangxi in 1969, the ritual eating of livers of slain class enemies?"

If the CCP learned one lesson from the fall of the Soviet Union—judges Liu Tong, a professor of history in Shanghai—it was that the Party can never surrender its control over how history is written. "The destruction of history," says the professor—by which he means the officially dictated version of history—"would be the first step toward the destruction of the Party. It would shake the faith of the people."[67]

The acts of cannibalism referred to by Perry Link were revealed in Zheng Yi's 1993 work *Scarlet Memorial*.[68] That book, and the ground-breaking *Tombstone* by journalist Yang Jisheng, were published in Hong Kong and Taiwan, and both are read in the USA and Europe. In China itself, they're banned.

Yang Jisheng worked for the remarkable magazine *Yanhuang Chunqiu*. This was no dissident publication; on the contrary, it was founded in 1991 by old Party veterans, though they had learned their lessons from Mao's tyranny. The editors were almost all CCP members, whose aim was, as Deng Xiaoping once instructed his people, simply to "seek truth from facts." The things the magazine brought to light exposed some of the CCP propaganda's favorite stories as fairy tales. It is just this kind of genuine truth-seeking that Xi Jinping calls "historical nihilism": it shakes the faith that is the privilege of the blind and the deaf.

Yanhuang Chunqiu was another victim of Xi's campaign against any attempt to revisit the past, and in 2017 propaganda officials took possession of the magazine in a hostile takeover. Du Daozheng, the aging founder and editor-in-chief, had just been taken to the hospital. From his sickbed, the 92-year-old condemned the takeover as an act "from the days of the Cultural Revolution." Earlier, in 2011, Du Daozheng had compared the situation in China to a pressure cooker: "The tighter you screw the vent, the higher the pressure grows inside, and one day the whole thing will explode." The following year, Xi Jinping took over the Party leadership, and since then the vent has been screwed one turn tighter every day.

The amnesia machine doesn't just erase the Party's great historical lies and crimes; it also works tirelessly on a small scale, day after day, month after month, year after year, burying all the small and medium-sized missteps, slip-ups, accidents, and catastrophes that collectively might make people wonder if there could be something rotten in the system. Events on which reflection and analysis might find a foothold simply evaporate, become things that never happened. Spaces in which connections and patterns might become visible are closed off. Often, all that remains is a quiet echo: was there something there? What memory the people still have is populated by shadows and outlines.

The AIDS villages in Henan Province are one example. Unscrupulous Party cadres and business-people worked hand in hand to build up lucrative blood-donation networks in rural areas. Thanks to the operators' greed and negligence, these were run so unhygienically and with such scant regard for medical protocols that eventually whole villages caught HIV and waited for death together.

Or the explosion of a storage facility for toxic chemicals in Tianjin in August 2015. Toward midnight, a column of fire taller than a skyscraper shot up just a few hundred meters from an expensive residential area. An inferno raged in the heart of one of China's wealthiest and most modern cities, two hours' drive from the capital. And all because of an illegal warehouse for dangerous substances, containing deadly chemicals the authorities had known nothing about. A clueless local authority, corrupt officials, hopeless leaders who dispatched dozens of firefighters to certain death from the poisons and the flames. The devastating result: 200 dead and missing in the blaze; a crater in the middle of the city; contaminated ground and buildings. The people were shocked. But they only learned what had happened from the early images of this "hell on earth" that circulated on social media. For the first few days, the Party apparatus seemed to be paralyzed.

Not long after the disaster, I got a call from an advertising executive friend of mine. He and his wife had decided to emigrate. "We've had enough," he said. "It was only coincidence that we heard about Tianjin this time, coincidence that everything came to light this time: how corrupt they are, how they hush everything up, how they cover for each other, how little value they place on human lives. But how many equally terrible things do you think are going on around us right now, about which we'll never hear anything?" He told me they were now looking for schools in Europe for their seven-year-old son. "What a country this China is. You're never safe, and there's no hope. You know, we

Chinese can swallow anything. That's the way we are. But we have to get our son out of this country, at least, and make another life possible for him."

The Party knew what was at stake. They knew that this accident could be traced back to the system's lack of transparency, corruption, and incompetence. Censorship and propaganda struggled back to their feet. Their first job was to turn stories of sadness and mourning into news of heroic deeds and celebration of all the brave soldiers, firefighters, and nurses sacrificing themselves in the service of the Party and the people. And then very quickly—in the space of just a few weeks—Tianjin vanished from the news altogether and was never mentioned again.

In the UK, such an event—one thinks of Hillsborough in 1989 or Grenfell Tower in 2017—would have prompted years of analysis, research into its causes, mourning, and annual remembrance; but in China, it melted away into nothing. With the devastating earthquake in Sichuan on May 12, 2008, the government attempted an even more spectacular conjuring trick. 69,000 people lost their lives in that earthquake, including 5,000 schoolchildren buried under the rubble of their schools. The schools had clearly been thrown together using low-quality materials—meaning that many of those children were as much victims of corruption as of natural disaster. It was impossible to cover up this fact, and initially it caused outrage across the country.

Yet in 2018, the Wenchuan regional government declared that the tenth anniversary of the earthquake should not be a day of memorializing—instead, it advertised it as a "day of gratitude." On this day, a euphoric article in the state press proclaimed, the earthquake victims would be feeling "bubbling springs of gratitude" for all the "beautiful, clean" new buildings and the "great love" that country and Party had shown them following the catastrophe.[69] Not a word about all the suffering, the bitterness, and the accusations from parents who had lost their children, some of whom protested so long and so loud about the lies and cover-ups in 2008 that they were eventually arrested and silenced.

"Gradually we become accustomed to amnesia and we question people who ask questions," writes Yan Lianke. "Gradually we lose our memories of what happened to our nation in the past, then we lose the sense of what's happening in our nation at present, and, finally, we run the risk of losing memories about ourselves, about our childhood, our love, our happiness and pain." And so China is silent about those who starved to death during the Great Leap Forward, and silent about the roots of the Cultural Revolution, and silent about the massacre of its children in 1989 in the heart of its capital. But beneath this leaden silence, a rancid brew of pain, guilt, and bitterness is fermenting, sending poisonous bubbles rising to the surface of today's China. One of the most toxic of these is a kind of nostalgia without memory, and thus without truth. A nostalgia dreaming the monster of the past into a thing of beauty—and wanting it back.

In summer 2012, there was another fight with Japan over the Diaoyutai Islands in the East China Sea. Something unheard of under the CCP took shape in the capital: a demonstration with tens of thousands of participants. The demonstrators were bussed in to give free expression to the people's anger with Japan. And for the first time since the end of the Cultural Revolution, the mob that moved through the streets of Beijing was not merely wishing an ugly death on its enemies (some of the placards said "Wipe out Tokyo!")—alongside the hate-filled slogans, people were once again brandishing Mao placards. "Come back, Mao," they said.

The year before, Bo Xilai, charismatic governor of Chongqing, had tapped into the same nostalgic spirit for the 90th anniversary celebrations of the birth of the CCP. Attracting admiration across the country, the maverick party boss—who would soon be deposed by his arch-rival, Xi Jinping—had gotten crowds in his city's parks singing the "Red Songs" from the Mao era again: "The love for mother and father does not compare to the love for Mao Zedong." The clique of intellectual New Left members and neo-Maoists, who had given Bo Xilai their enthusiastic approval and hailed him as China's great hope, is still one of the noisiest movements on China's internet—and many of them have now transferred their hopes to Xi Jinping. These are people who say that the millions who starved to death during the Great Leap Forward were a myth invented by the CIA.

A few steadfast souls remain horrified by the amnesia. Among them is the former Red Guard soldier Zhang Hongbing (see chapter one), now a lawyer, whom I interviewed at his home in Guzhen, a small county in the southern province of Anhui. He is convinced that "power [must be] caged," and the truth brought out into the light. This is a battle to save the children, he says: the children of China, who are once again so naïve, so easy to lead astray. Zhang Hongbing was sixteen when he handed his mother over to the executioners. Just because, at dinner one night, she had said she'd rather see someone other than Mao as head of state.

"But child," she said in response to his outraged protests, "you have no idea about class struggle."

The son leaped up. "Who is your child here?" he cried. "We are the Red Guard of Mao Zedong. If you go on spewing poison, I'll smash your dog skull!"

When his mother also lost her temper and said fine, then she was going to rip the picture of the great Mao down off the wall, his father chimed in against his wife: "Fang Zhongmou! You are an irredeemable counter-revolutionary and from now on, you are not a member of this family. You are the enemy. We will fight you!" The son denounced his mother to the Revolutionary Committee, writing that his mother deserved the death penalty. A few weeks later the revolutionaries fulfilled his wish and executed her.

China's schools, according to Zhang Hongbing, bore a great responsibility for what happened in those times. "They are still turning out devoted subjects today," he says. "Slaves. Children raised by wolves." People like Zhang not only receive regular "invitations for tea" from concerned police officers, they also receive angry letters from the general public. "What's your game?" "Our lives are so good today. Why go dragging up the past?" "Do we not live in a 'harmonious society' today? You're trying to destroy the harmony!" Zhang sighs: "Why do the sons and daughters of the Chinese people understand so little?"

The Party is so good at erasing history that it sometimes trips itself up. In 2007, a civil rights activist in Chengdu named Chen Yunfei succeeded in placing a small ad in the paper as a "tribute to the strong mothers of the victims of June 4." When the young editor asked what that June 4 date signified, she replied: "A mining accident," and he was quite satisfied. Afterward, he and two colleagues were fired—although their ignorance was merely evidence of how successful the efforts of state censorship had been.

I was reminded of this when I went to see a modern dance production in the heart of Beijing, given by the Taiwanese Cloud Gate Dance Theatre at China's National Theatre, and using the "Nine Songs" of the poet Qu Yuan written more than 2,000 years ago. The choreographer Lin Hwai-min had transposed the "Nine Songs" to Taiwan, and in the process brought some of them into the modern era.

Even so, I was unprepared for the final scene. The stage was dark. Very slowly, the shadow of a tank emerged from the back of the stage. Dancers stood up, just a few at first, then more and more; they came together to form an army of upright shadows, marching silently and solemnly toward the tank. Then suddenly: machine-gun fire. The dancers stumbled, twitched, fell to the floor.

Lin Hwai-min had created the scene as an homage

to the victims of the dictator Chiang Kai-shek (1887–1975) and his White Terror; for him, it was an homage to the victims of the massacre that took place in Taiwan on February 28, 1947. But how many of the audience in Beijing knew that? And how was it possible, at this moment, in this place, just 200 meters as the crow flies from Tiananmen Square, not to see the images as an allegory of June 4, 1989? I sat there for a while, thunderstruck.

How on earth had this slipped past the censor? Opened in 2007, the National Theatre is a silver half-globe floating on water, devised as an architectural statement that China is part of the modern world. Had all memories of 1989 been obliterated from the heads of those responsible, even here? It was a bitter irony that only the Party's brainwashing had allowed the victims of 1989 to be commemorated just a stone's throw from Tiananmen Square. Their shadows returned, if only for an evening.

Eventually, the shadows were all mown down and lay on the stage. Silence. From the wings, out of the darkness, other dancers entered, carrying little lights, which they placed on the stage one after another until in the end, they formed a sea of stars.

The Mandate
from Heaven

How the Party Elected an Emperor

"One-party dictatorship always leads into perdition."

Xinhua Daily, *mouthpiece of the Chinese Communist
Party (then in opposition), in 1946, writing
about the dictatorship of the nationalist Kuomintang*

China has a government, just like we do. Except in China, the government doesn't govern. Only the Communist Party governs. Ministers and ministries exist to carry out decisions made in the upper echelons of the Party. Sometimes not even that; they're mere decoration. And yes, of course we can talk about "President Xi Jinping," but that is Xi's least important title. It isn't the office of president that gives him his power; it's the office of general secretary of the CCP.

The CCP was founded in 1921, by a dozen secret communists who had fled from Shanghai to continue their meeting on a tourist boat, on South Lake in Jiaxing. Today, the Party has 89 million members, more than the whole population of Germany. Even after it seized power in 1949, it never entirely stopped acting like an underground organization. To this day, the words of the sinologist Simon Leys remain valid: he once described the daily task of the Party-watcher as "the art of interpreting non-existent inscriptions written in invisible ink on a blank page."[70] And when Richard MacGregor's *The Party* (a standard work on China's Communist Party) came out in 2010, its subtitle was: "The Secret World of China's Communist Rulers." The most memorable quote in the book came from a professor at the People's University in Beijing: "The Party is like God. He is everywhere. You just can't see him."[71]

The Party is still like God, and it is still everywhere—but one thing at least has changed with the arrival of Xi Jinping: you can see it again. Xi is making sure the Party is worshipped, openly and fervently. After decades of the "reform and opening-up policy" in which the CCP often ducked out of the limelight and contented itself with pulling strings behind the scenes, Xi decided it was no longer in the Party's interests to hide or water down the essence of its system: "It doesn't matter whether it is the government, the military, the people, or the schools; east, west, north, south, or the center—the Party rules everything."[72] Not much room for ambiguity there: Xi is doing away with the separation of Party and state that Deng Xiaoping once envisaged. Away with the freedoms for which civil society and parts of the media had fought. Away with the illusion that the People's Liberation Army serves the country and the people first and foremost. Strictly speaking, China's army has never been China's army—it was always the Party's army, and it still is today. "You should have unshakable loyalty to the Party's absolute command over the army, always follow the Party's call, always obey the Party," Xi Jinping told the troops who had assembled for maneuvers to mark the PLA's 90th anniversary. "Wherever the Party points, you should march."

Party cells are also regaining power in universities, think tanks, NGOs, and companies; the Beijing historian Zhang Lifan speaks of the "partyfication" of the country. China's colleges, he says, are in the process of transforming themselves into "Party schools." Xi Jinping is a control freak: he is using the reanimated Party cells to recentralize power, and he is making sure people get the message. State, people, and Party—for the leader and mastermind of the CCP, these are all just different manifestations of the same thing, and, like the Holy Trinity, they're difficult to separate. If the Party is now swallowing up many parts of the state and taking its place, then it is just making the system a little more honest. The restructuring of the government in spring 2018 was a logical step: CCP steering committees took control of the ministry of the economy, the finance sector, foreign policy, and cyberspace, which had been within the purview of government authorities. The Party's central propaganda authority took over the regulation of film, press, and publishing from state authorities.[73] And the CCP's United Front Work Department swallowed up the state offices for religious affairs and for relationships with Chinese citizens abroad.

The Party is no longer hiding. And businesses in China are starting to feel it, too. There have always been Party cells within companies, though for the past few decades they've been nothing more than a formality. Sleeper cells. Now they are being woken up; they want to have a voice, to help make decisions. Even in private companies; even in foreign companies. The Chinese joint-venture partners of Western firms are demanding that the Party cells in their businesses are included in strategic business decisions.[74] All Party members, so it says in the statutes of the CCP, must place "the interests of the Party and the people above all else." One Chinese judge put it this way: where there is a conflict of interests, the "Party nature" of a CCP member must always outweigh his "human nature."[75]

At the offices of the internet firm Tencent in Shenzhen, there is a board on the wall showing how many employees are Party members (in 2018, that number was 8,000). The Tencent penguin on the board proudly displays a hammer and sickle on his chest.

Private companies have long been the engine of China's development and modernization: when Xi Jinping took power in 2012, they represented half of all investments in China and three-quarters of production. This applies particularly to the areas of information technology: all of China's internet giants and almost all start-ups in artificial intelligence are private companies. But China's new entrepreneurial class knows that it is at the mercy and arbitrariness of the CCP. Only those entrepreneurs who are constantly aware of their delicate role in the party's realm are successful. "There is no such thing as free enterprise in China," says the American professor of economics Christopher Balding, who taught for a couple of years in the southern Chinese city of Shenzhen. "It is all just varying degrees of Beijing servitude."[76] China's businesspeople are not supposed to simply earn money.

First and foremost, as Xi Jinping himself once told a group of business leaders, they have to "love the fatherland, the people, and the Communist Party and actively practice the core socialist values." In May 2018 the Cyberspace Administration and China's internet giants came together to launch the China Federation of Internet Societies (CFIS),[77] whose stated purpose is to "promote the development of Party organizations in the industry." Afterward, the *People's Daily* said that the CFIS would "conscientiously study and implement the spirit of Xi Jinping's Strategic Thought on [Building] an Internet Power."[78] The vice presidents of this new organization are China's most famous internet bosses, Pony Ma (Tencent), Jack Ma (Alibaba), and Robin Li (Baidu), whose "voluntary" cooperation was highlighted.

The CCP opened its ranks to "new layers of society," as it called them, more than 15 years ago. Since then, the relationships between power and money have been particularly close. In March 2018, when the National People's Congress and the Political Consultative Conference (which always meet in parallel) came together to hold another coronation mass for the "people's leader" and "helmsman" Xi Jinping, no fewer than 153 of the delegates were also listed among China's "super-rich" in the Hurun Report published in Shanghai. Together, their massed wealth was the equivalent of $650 billion, or just slightly less than the GDP of Switzerland.

The big cheeses in the online sector vie with each other to sing the praises of the Party and its leader. Liu Qiangdong, head of the e-commerce giant JD.com, shared a surprising epiphany with his audience in August 2017: he had been looking at the progress of artificial intelligence, he said, and had "suddenly discovered that communism can actually be realized in our generation." In the autumn of the same year, Alibaba boss Jack Ma presented his insights from the CCP's recently-held 19th Party Congress. In summary: over the last five years, the Party had grown "even greater in its ability to improve and reinvent itself."[79] Private companies have never had it so good as they do in today's China. Jack Ma pointed to permanent political disagreement in Washington and praised China's "advantages": no other country in the world could boast such "political stability." This subservience is a calculated strategy in a country where it is the Party, and not the market, that makes the rules. A Party whose inspectors and agents regularly make the bosses of even the largest firms vanish overnight, and keep them for days, weeks, or months while they "assist with investigations." Incidentally, Alibaba's

CEO Jack Ma was revealed by the *People's Daily* to be a long-standing Party member in November 2018. Liang Wengen, the billionaire and head of the heavy-equipment manufacturer Sany, once publicly vowed: "My property and my life belong to the party." And he is right: if the CCP wants to take either or both, he is powerless to stop them.[80]

The Party itself often inadvertently lets slip that the all-around renewal of the CCP under Xi still has a long way to go. In the city of Ningbo, for example, functionaries printed their own handbook to encourage "a more noble spirit" among cadres.[81] One chapter is headed: "What One Should Not Say to the Masses," and lists 44 examples that, one may infer, are in common use. Not every Party member, it seems, is working his fingers to the bone in the service of the people. Here are seven of those things not to be said:

"I completely agree with you. But the officials up there in the city don't."

"I'm busy, don't disturb me."

"If something happens, don't come to me under any circumstances. I don't want to be promoted, anyway."

"So much work and so little money. I'll do it tomorrow."

"Don't talk to me about anti-corruption
   regulations . . . They've been doing that for
   years. We've gotten through two years now, it
   will pass, everything will be all right in the end."
"The Party is just as bad as you say."
"If there's an election, vote for me. I'll make it
   worth your while."

Immediately after Xi Jinping's speech at the 19th Party Congress, the internet giant Tencent, which is listed on the Hong Kong stock exchange, brought out the app "Excellent Speech: Clap for Xi Jinping." It was a game in which the player had 19 seconds to applaud Xi Jinping as enthusiastically as they could. And it was a hit: just 24 hours after its release, people had already given Xi a total of more than a billion claps.

Applauding Xi Jinping has become a national sport. Shortly after the Party Congress, the state network CCTV began its evening news program with a memorable four minutes of uninterrupted applause for Xi from worthy comrades. As the Australian sinologist Geremie Barmé observed, Xi was no longer just the "Chairman-of-Everything and Chairman-for-Life"; he was now also "Chairman of Everyone and Chairman of Everywhere."[82] The Party had enshrined him alongside its highest thinkers, by including him in the Party statutes with "Xi Jinping Thought on Socialism with Chinese Characteristics for a New Era." A few months later, "Xi Jinping Thought" was added to China's national constitution, making Xi the first leader since Mao to be given this honor during his own lifetime. In practice, it means that for the rest of his life, he is unassailable. From now on, anyone who criticizes him automatically makes himself an enemy of the Party and the constitution.

Many have seen the writing on the wall. In Yugan County, Jiangxi, a province which is home to many Christians, officials have forced people to remove pictures of Jesus from their houses and replace them with portraits of Xi. In the same province, in 2019, a Catholic church in the city of Ji'an was forced to replace the image of the Virgin Mary with a portrait of Xi. Or in Henan Province, where local Party leaders who made a pilgrimage to Lankao County in the northeast of the country to visit a tree. More precisely, an empress tree—one that Xi had planted with his own hands, eight years previously. The photos could be admired on the Party cell's website, where you could also read about the group gazing upon the tree in meditative awe, while reflecting "carefully on the mission of the Communist Party." They then listened to a reading of a poem from the pen of Xi Jinping: its verses pay homage to a good Party functionary who, just before the Cultural Revolution, worked himself to death in the service of the people.

In recent years, nothing has been more guaranteed to provoke an allergic reaction from Party representatives than the suggestion that a new personality cult is being born. "No, certainly not!" exclaimed Xie Chuntao, one of the leaders of the Central Party School, shortly after the Party conference. China had learned from history, he said, alluding to Mao. "Such a thing will not happen again." What we were seeing was the people's "respect and love" for Xi—both of which were "natural" and came spontaneously "from the heart."

Mao's disciples used to say the same thing. And where once upon a time the propaganda department invented the Little Red Book for Mao, it has now given Xi a Little Red App. Its name is "Study (*Xi*) Strong Country"* (*Xuexi Qiangguo*)—and, as was only right and proper, it became the country's most-downloaded app shortly after its release in January 2019, ahead of WeChat and the wildly popular video app TikTok.

---

* In fact, the app's name, "Study (Xi) Strong Country," is a pun—*Xuexi* is the word for "study" but also contains the president's name, suggesting users are to "study Xi." Such are the wonders and ambiguities of the Chinese language that it can be read in several ways.

Party members can not only find the president's latest speeches and thoughts there; they can peruse the Party press, download Marxist classics and revolutionary films, chat, and send each other "red envelopes" containing gifts of money. But most important—and this is what makes it a killer app—they can collect points: one for every essay they read from the pen of Xi, and one for every video they watch.

There are ten points available for every 30 minutes of "Xi-time" (as one heading in the app calls it) you indulge in, or for giving correct answers in a Xi-themed quiz. And if you give the app your attention between 6 a.m. and 8:30 a.m., or in the evenings from eight o'clock onward, the points you earn are automatically doubled. Gone are the days when Party members could let the evening news wash over them and put the *People's Daily* in the recycling bin unread; now your smartphone registers every minute you devote to something and every paragraph you read. The Party is creeping back into every crack of people's lives. Party cells in some parts of the country have already started holding Xi study competitions with the app, and giving their workers point targets to reach, and keeping some people awake half the night. The CCP propaganda department also stated that journalists from important Party and state media organizations would receive their press cards only after a test of their ideological stability via the app. In autumn 2019, the Berlin cybersecurity company Cure53 published a study, together with experts from the Open Technology Fund. It revealed that the app was designed as a spying tool and, via a "back door," not only allows the authorities to view all user files, downloaded apps, and other information on Android mobile phones but also gives the CCP the opportunity to, for example, alter those files or install keyloggers, software that logs every keyboard input by the user. According to the study, skimmed information such as connection and location data is transmitted daily to the xuexi.cn domain, which is managed by Alibaba. Alibaba also programmed the app for the party.[83]

Xi was once lauded as the "core" of the leadership team. Those days are gone. He is now the one and only, the "reform strategist," the "supreme commander," the "world leader," and the "helmsman." The Party press is beside itself with amazement and awe. "What is it that makes the great Xi Jinping Thought so great?" was the headline in one issue of *Study Time*, a Party School newspaper.

The amazement is warranted. Viewed from a distance, the sheer speed, the tactical skill, and the force with which this man has ascended to the emperor's throne is incredible. It couldn't have been foreseen in

2012, when the Party, in desperate need of a savior, elected a bland apparatchik by the name of Xi Jinping. And it certainly can't have been what all sections of the ruling kleptocracy wanted.

Xi is a member of the red aristocracy, the son of an old revolutionary. As a young man, he was sent off to the countryside, and for years he lived on the Loess Plateau in central China, later climbing the career ladder as a bureaucrat in the provinces. In all that time, he made no great impression. But he was doing what you have to do to survive in an apparatus as merciless as the CCP: he kept a low profile. This probably explains why he was raised to the top of China's hierarchy: powerful Party grandees thought this colorless man would pose no danger to them.

Xi Jinping surprised them all. He proved to be a superb power strategist, dispatching his rivals one after another—even those who had previously seemed invincible. He was aided in this by the longest-lasting anti-corruption campaign the People's Republic had ever seen. Shock and awe. Gone were the days when such political campaigns would come and go like the tides, when anyone with real power could get themselves to safe ground in time. Party bigwigs have always removed political power from defeated opponents, but only rarely did they take their money and more rarely still, their freedom. Xi broke with this tradition. He took everything from his enemies. He has drawn all power to himself, bringing the security services and the army under his control. On the global stage, too, he has proved an astute tactician, quick to exploit other countries' weaknesses: for instance, with the speech he made to the World Economic Forum in Davos just after the world-blind Donald Trump was elected US president, in which he promoted China's leading role.

Of course, Xi has allies in the CCP, chief among them the current vice president, Wang Qishan. He was Xi's right-hand man during his first five years in office, when he was the Party's most senior anti-corruption crusader. Xi and his allies have resurrected the strong leader, no doubt partly to compensate for the fact that the shared values that are supposed to cement society together are in short supply in China. Sentiment and rhetoric are being made to stand in for genuine reform. Yes, Xi continues to promise the poor and the middle classes a better life—but because he is aware of the great dissatisfaction in the country and the sheer numbers who have been short-changed, he also gives his people the prospect of sharing in the Chinese dream of a strong nation.

The aim of the personality cult around Xi is to marshal the people behind a symbol. In part, the China

Dream, the growing nationalism, and the rhetoric of unity are the Party's response to the extreme split in Chinese society between rich and poor, the wealthy cities and the left-behind interior. Economic growth is slowing year on year. This may not concern many members of that small elite who have amassed obscene wealth with the help of the CCP, but ordinary people are worried. Now, though, the Party has Xi the problem-solver, who doesn't forget the little people. Xi the visionary, who gives China a dream. Xi the hero, who makes the nation a force on the world stage again. Xi the thinker, whose volume of essays about government naturally just happens to be lying on Mark Zuckerberg's desk at Facebook HQ when Chinese reporters go past. The book, so the *People's Daily* reports, is "thrilling readers" from Cambodia to the UK. Xi the wise ruler, who reconciled the ancient philosophers Confucius and Han Feizi first with each other, and then with Marx and Mao, thereby proving to China its "exceptional status" among the nations of the world (exceptional meaning that China, alone, is made for the eternal rule of the Communist Party).

Again and again, the propaganda machine shows images from the hard years of Xi's youth, when he was dispatched into the countryside like so many of his generation, to "learn from the peasants." We hear about him working tirelessly to help the farmers of the Loess Plateau, in the yellow earth of Shaanxi Province, the mythical heartland of Chinese civilization; sleeping in caves, looking after sheep, and "carrying 100 kilos of wheat up a five-kilometer mountain path, without changing shoulders even once." The village's name is Liangjiahe, and it is destined to go down in history. China's academics have been called upon to conduct research on the "great teachings of Liangjiahe." On the evening news, you could watch Xi in the place where he once worked, patting babies, admiring apple trees, and saying, "This is where I left my heart." He asks the farmers whether their life is good. They reply: "The Party's policy is good. The farmers are full of hope."

The message is: "Look, here is someone who comes from the people himself." In reality, Xi is one of the so-called "princelings," a scion of the red aristocracy, son of Xi Zhongxun, a revolutionary once close to Mao and Deng. He is one of the chosen ones, and he knows it. Xi's self-confidence comes from his pedigree. All his life, he's been groomed for the task he is now fulfilling: the salvation of the CCP's power in a hostile environment, in a world for which it was not really made.

Xi is aware of his country's heritage, and he makes deliberate use of it. When he brought an end to the extravagant binges that corrupt party functionaries were enjoying on a daily basis, it was with a decree that henceforth, only "four dishes and one soup" would be served at official meals. No more of those sprawling banquets with which government officials all over the country had been filling their days. It is hard to believe Xi didn't know that the exact same menu—"four dishes and one soup"—had been prescribed 650 years previously by the first emperor of the Ming dynasty, Zhu Yuanzhang, for his courtiers in Nanjing. Zhu had fought his way to the top from lowly beginnings in a poor peasant village, and won the mandate from heaven that made the emperor into an emperor.

In China, heaven traditionally bestows its mandate on a person when he is a particularly virtuous and capable leader. A person with whom people align themselves like iron filings around a magnet; someone who can bring order to the world by the sheer force of his charisma. And if you don't really believe in heaven, but are confident that you are the most virtuous and capable of all leaders, then you can simply seize that mandate for yourself. When Xi Jinping appeared at a press conference after the 19th Party Congress, he referred to a famous Yuan dynasty poem. Just as a plum blossom did not need anyone to shower it with praise, and it was content that the universe was filled with its scent, he said, "I don't need cheap compliments from anyone.

It is enough that my integrity fills the universe." These are the words of someone already sure of his imperial status and charisma.

A few days earlier, I had attended an event in Beijing's exhibition center which, according to the accompanying PR material, was all about showcasing the country's progress. In reality, it was a High Mass for the man whom the *Economist*, the central mouthpiece of capitalism, had that same week proclaimed the "most powerful man in the world." Pride of place was given to a huge tower of bookshelves, filled with titles by Xi Jinping. An old woman was leafing through his *The Governance of China*. She introduced herself as Mrs. Liu, 72 years old, a practicing Buddhist. Oh, I said, a Buddhist? Yes, Mrs. Liu replied: after all, people need religion—especially in China. And that, she added, was also why she revered Xi Jinping so greatly. "He's not an ordinary person. He has a destiny." As she saw it, the Party intended to bring liberation to the country and peace to people's souls. "You know," she said, "the Communist Party is a religion, too. It's good for us."

The Dream

How Karl Marx and Confucius Are Being Resurrected, Hand in Hand with the Great Nation

"Yeah. You. You. You know / Communism is sweet like honey / I'm your Bruno Mars / You're my Venus, my dear Karl Marx."

Rap song by the band Xiangshui zuhe ("Perfume"), 2016

Marx may be dead, but on May 5, 2018, he managed to celebrate his 200th birthday in grand style, with a Sino-German festival in his hometown of Trier. In China, his rebirth is old news: the country has made it its mission to become the "global center for Marxism research." (It's also the global center for Karl Marx worship—not that there's much competition.)

All this is thanks to Xi Jinping, who took the busts of Marx down from the shelf, where—even in China— they were gathering dust, polished them up and invited the Party and the people to kneel before them once more.

New statues of Marx were sculpted and cast in bronze. One found its way to Trier, a gift from the People's Republic of China to mark the bicentennial of the city's great son. It stands 4.4 meters high—to the relief of Trier's citizens, almost two meters shorter than originally planned, but even so a colossus. As is only right and proper, according to Wu Weishan, the sculptor: big philosopher, big idea, big China. Wu was enthusiastic in his praise for little Trier: "Home of Marx, dreamland of our spirit!" he wrote in an article for the *People's Daily.* As curator of the National Museum, the sculptor is a cultural bureaucrat, though for many years he has also been kept busy producing works commissioned by the Party.

His Marx—marching forward with determination in a billowing frock coat—symbolizes "China's faith in its own theory, its own path, its own system, and its own culture." As befits a state artist in China, Wu was simply transcribing the exact words of his helmsman Xi Jinping, who demanded these "four forms of self-confidence" from his people. What they mean is that

China is self-sufficient now, and so sure of itself that it is setting out to hijack Marxism and reinvent it in Chinese. Xi Jinping Thought is the new Marxism, says the state press. The adoption of Marxism by the CCP was "totally correct," said Xi in his speech on Karl Marx's birthday, and "the sinification and modernization of Marxism" was now also "totally correct."[84] He urged his comrades and the whole Chinese nation to make aligning themselves with Marx "a way of life."

In 2012, Xi Jinping took over a country riddled with corruption and a Party showing all the signs of internal disintegration. There was great uncertainty throughout the country, reaching deep into the Party ranks. From the start, Xi Jinping had a clear battle plan to take back control of Party, state, and society. Central to this plan was a wave of ideological purges, which washed through all the country's important institutions in turn—and was intended to rinse away anything with a whiff of "Western values."

While Xi was still in the process of acquiring a stranglehold of censorship on social media, a document from the CCP Central Committee, which was to become famous as "Document Number Nine," began circulating among Party functionaries. Its real title was "Communiqué on the Current State of the Ideological Sphere." This report from April 2013 is a head-on attack against what political thinkers and civil rights activists the world over describe as "universal values." For the authors of the document, though, they are without exception "Western values," "Western principles," "Western ideas." The Party regards all these ideas as subversive, feels threatened by them, and so is at pains to brand them as unsuitable for the world beyond the West.

The document calls on Party members across the country to take up the struggle against concepts such as civil society, the separation of powers, an independent judiciary, independent media, and coming to terms with one's own history. All these ideas had taken root in the three decades before Xi Jinping took office, even within the Party itself—at least among its more open-minded and critical figures. Now the West was an ideological opponent once more.

The campaign quickly gathered pace. The propaganda apparatus systematically took various groups and institutions to task. One of the first to attract its attention was the Chinese Academy of Social Sciences (CASS), the government's most important think-tank, which in recent years had produced many original and independent thinkers. In June 2014, a senior cadre from the powerful Central Commission for Discipline Inspection—the same organization that for a long time had led the fight against corruption—accused CASS of having "ideological problems"; its infiltration by "foreign influences" had become too dangerous. A few days later, the academy dutifully announced that from now on, it would make checking the ideological purity of its researchers a priority. In December of the same year, the CASS president Wang Weiguang went even further, writing that "the class struggle" lived on in China—a return to Maoist jargon that alarmed many people.

One month later it was the turn of members of the CCP itself: societal change over the past three decades had unfortunately led to "a loss of faith and moral decline" among many functionaries, according to a declaration that revealed a great deal about the Party leadership's worries and the subversive attraction of the ideas they had demonized. The cadres had to re-immunize themselves with Marxism, it said, "so that they do not lose their way in striving for Western democracy, universal values, and civil society."

In September, China's journalists were next in line, receiving the order to follow only "Marxist news values." And in October it was the artists, when Xi Jinping told them in no uncertain terms that from now on, they must once again create only works that "serve the people and socialism." Both the jargon and the main thrust of his speech had echoes of the notorious address given by Mao Zedong in 1942 that would gag China's artists for decades.

Very quickly, the state apparatus also set its sights on the country's universities. In December 2013, Xi had already made a speech calling for Chinese higher education institutions to take a new ideological direction, and it took them less than six weeks to obey. Their institutions now formed the "front line" on the ideological battlefield, education minister Yuan Guiren told the vice chancellors. The specific enemies they had to wipe out were Western values and Western textbooks. The university vice chancellors and Party secretaries instigated a purge of the course plans and lecture halls. In place of Western ideas, the minister said, they must use all their strength to drive the ideas of Marx and Xi Jinping "into students' classrooms and into their heads." (This prompted a few brave souls who still dared to ask questions in China to wonder out loud from which point of the compass Marx and socialism had come in the first place.) In *Seeking Truth*, the magazine of the Central Party School that sets out the ideological direction in which the CCP will march, Minister Yuan upped the ante: young teachers and students, he wrote, were "the primary targets of infiltration by hostile foreign elements." The ministry announced the introduction of new textbooks for schools and universities, which would implant "red genes" back into students.

It was only natural that this ideological zeal would hit the universities harder than most institutions: they are the home of a generation of professors and lecturers who have seen more of the world than almost any other group in China. As teachers, they also have a great influence on the brightest and best of the young generation. The attacks have not yet subsided, and the purges are ongoing. Inspectors from the Central Discipline Inspection Commission have toured the country's universities several times, criticizing even elite institutions such as Tsinghua and Peking Universities for "weak ideological and political work" and demanding that they submit immediately and completely to "the Party's guidance."

As a rule, such criticism from the feared Discipline Commission results in panicked activism from the institution criticized. More than a few independent minds have been fired or sidelined, perhaps for permitting themselves a critical remark about Mao Zedong, or showing a shade too much enthusiasm about free speech. Others have disappeared behind a veil of silence. Even in 2012, we correspondents could still find some of our most exciting and original sources at CASS or the major universities; and they were often

Party members. Today, they often don't even pick up the phone.

The informer is back. "Student information officers" are again in large numbers in the lecture halls and report to university Party cells about every word of the faculty that deviates from the Party line. But the Party doesn't just use threats; it also buys the loyalty of university teachers. Two friends of mine, both English lecturers at a large Beijing university, informed me with great astonishment about the introduction of a new *weiwenfei*, the "stability maintenance" being paid to all teaching staff—provided that they hadn't attracted negative attention by making critical comments in their classrooms. One had received 40,000 Yuan and the other 50,000, sums which at the time were the equivalent of around $5,640 and $7,050 respectively. "Just imagine," said one of them. "All that money!"

Peking University—once a beacon of liberal values in China—led the way: in 2015, it named one of its buildings after Karl Marx, and put together a team to collect Marxist classics. The project was called *Ma Zang*, which can be translated as "Marxist canon." This is not without irony: until then, *zang* has been the term used by Chinese scholars for collections of the sacred or canonical texts of Buddhism, Taoism, or Confucianism. When the Beijing *Global Times* reported on the project in 2015, the article's author allowed himself a note of skepticism. On the political education courses regarded by many students as a hateful, tedious hoop to be jumped through, he wrote: "Marxism itself has lost attraction among students. The teachers can do few things about it. The students learn Marxism just for academic credits."[85] Three years later, the same newspaper felt duty-bound to report: "More than 90 percent of Chinese college students surveyed said they are inspired by political and ideological courses in China."[86] The paper quoted the education minister, who proudly announced that students had to "fight for a seat" on some popular courses. Which sounds like something an education minister might say when he wants a pat on the head from the Party leadership.

The ideological training seems to have made a particularly deep impression on the administrators of the sperm bank at Peking University Third Hospital. When they were looking for new donors in 2018, they not only required potential applicants between the ages of 20 and 45 to be free of inherited or infectious diseases and "obvious male-pattern baldness"; the ad on Weibo also asked for "excellent ideological qualities." Sperm donors would only be accepted if they "love the fatherland, support the rule of the Communist Party, and are loyal to the Party's mission." So that the red genes really are passed on.

What does the CCP leadership intend to achieve with this ideological revival? A campaign like this only makes sense if the aim is to secure power at any price—including intellectual impoverishment. Yet the Party is striving to establish itself as the one entity with the intellectual capacity to govern judiciously, in order to achieve the ultimate goal of "the great rebirth of the Chinese people." With an eye on its big rival, the USA, the CCP wishes to propagate a world-class economy, world-class research, and world-class think-tanks. How it is going to achieve this while feeding intellectual pap to its best people remains a mystery. The truth is that the Party is desperate for original, creative, and critical minds in fields such as IT, physics, mathematics, and chemistry—but they must instantly fall into a deep sleep upon contact with politics, history, or religion.

Before Xi Jinping came to power, the Party had a problem: for several decades, it had lacked an ideology and a value system that would hold the nation together. "What we have in China is a total disconnect between theory and practice, where what is said is not only different from, but often the opposite of, what is done," the intellectual Rong Jian wrote in a much-admired essay in 2013. The massacre of 1989 had turned him from a researcher on Marxism into a curator. "Someone once told me that all China's problems stem from the fact that the capitalism that is being practiced is being preached as socialism. This paradox has persisted for a long time and has remained unresolved to this day."[87] Now Xi Jinping is attempting to fill the ideological vacuum that has existed in China since the death of real Marxism with Marxism's mummified remains. At first, it smacked of desperation. Unlike Mao Zedong's ideological purges—there was no shortage of the faithful in Mao's time—Xi's Marxism campaign has an overwhelmingly negative thrust. It defines itself first and foremost by its enemy: the West and its values. What it demands from the public is not faith, but gestures of submission to this stuffed Marx, gestures unifying the oath of loyalty to the Party with the rejection of non-Marxist Western aberrations.

When a handful of new genuine believers recently appeared in Xi Jinping's China, the Party's reaction was telling. These were young people, students, who called themselves fervent Marxists and were deeply affected by their country's inequality, the lack of a social welfare system, the scale of corruption and environmental damage, and above all the lot of the exploited working classes. When they got together to try to do something about these issues, they felt the fist of the Party. On August 24, 2018, alone, the police arrested 50 students in Shenzhen; not since Tiananmen Square in 1989 had the Party apparatus come down so hard on protesting students. To many, the spectacle of a supposedly communist party taking such repressive measures against Marxist students was further proof that Chinese socialism has become just an empty phrase.

Marx may be on Xi's lips, but Lenin is in his bones. "To dismiss the history of the Soviet Union and the Soviet Communist Party, to dismiss Lenin and Stalin," he said in an internal speech not long after he was named general secretary of the CCP, "is to engage in historic nihilism, and it confuses our thoughts and undermines the Party's organizations on all levels."[88] The Chinese journalist Gao Yu, who dared to bring both "Document Number Nine" and this secret speech to the public's attention, was arrested and sentenced to seven years' imprisonment in 2015 for "betraying state secrets." Following protests, especially from Germany and the USA, she was later released from prison "on medical grounds" and placed under house arrest.

The campaign against Western values brings forth some colorful flowers. The attack on Christmas is among them: over the last twenty years, celebrating Christmas has become popular in China's cities. As in other parts of Asia, young Chinese people in particular have adopted the more fun aspects of the festival, stripped of all religious components, to make it a time of year for parties and shopping. "Finally, a festival just for us, without any family stress, far away from parents and grandparents," a friend once told me. But the passionate official anti-West sentiment has not spared this Christmas Lite. In several provinces, the regional CCP leaders have instructed their members to avoid all Western festivals and customs. The ban also extends to Halloween, Valentine's Day, and even April Fools' Day, which the Xinhua News Agency warned against in the strongest terms on April 1, 2016, on its Weibo account. April Fools' jokes were "irreconcilable with core socialist values," it said. "So please do not start any rumors, do not believe them, and do not spread them!"

The main target, though, remains Christmas. In 2017, the city of Hengyang banned all civil servants and Party members from taking part in Christmas celebrations; after all, communism also means atheism. With one eye on the West's incursions into China more than a century ago, the Communist Youth League in Anhui called Christmas a "festival of humiliation," and the Pharmaceutical University in Shenyang warned its students against the "corrupting influence of Western religious culture." They would do better, it said, to work on restoring their "faith in their own culture." "Christmas is a festival celebrated by Christians and, is hence, a wound to the Chinese people. It is not suitable for the Chinese," said an article circulated on WeChat at the time. "We should not forget the shameful history of our country."[89]

Don't forget history: not such bad advice. In fact, if you follow it, you'll discover that the history of Christianity in China stretches further back than Christianity in America: the Nestorians were preaching the good news in the Tang capital Xi'an as early as the 7th century. The Jesuit missionary Matteo Ricci reached Macau in 1582 and Peking in 1601, and the presence of China's estimated 70 million practicing Christians today can be traced back to the seed planted by Ricci and his fellow Jesuits.

More generally, one could argue that Chinese people today have at least as much in common with us in the West as they do with their ancestors. They wear Western clothes and sport Western haircuts; their favorite cars are German and they live in a style of building once created by Western architects. They live by the Gregorian calendar, they prefer European classical music to their own traditional operas, they trust Western medicine more than traditional Chinese medicine, they write from left to right like we do, instead of vertically as they used to, and if they have the money, they send their children to Western universities.

Xi Jinping himself sent his daughter Xi Mingze to Harvard for four years. The Party, whose rule he would like to extend for eternity, was shaped by advisors from the Soviet Union; and the ideology he has begun to preach again was devised by a German philosopher. Concepts like democracy, freedom, equality, and the rights of the individual crept into China at the same time as Marxism, and have had a home in the country ever since. A century ago, they found a platform in the legendary periodical *New Youth*, which influenced a generation of reformers and revolutionaries, who subsequently turned China on its head. *New Youth* preached "science and democracy," and in his "rallying cry to the young," its founder Chen Duxiu railed against the "stale air" of old China and its traditions, stifling the nation and robbing it of its strength. A little while later, Chen became a Marxist, and in July 1921 he was one of the founding members of the Chinese Communist Party.

The modernizers' greatest enemy was Confucius. "Chinese culture is just a feast of human flesh, prepared for the rich and mighty," wrote Lu Xun, the greatest writer China had ever had, in 1925. Luckily for him, Lu died shortly before the Communist Party took power in 1949. The CCP and its leader swiftly co-opted the dead author, and he remains one of its saints to this day. This can mean one of two things: either they haven't read him, or they haven't understood him.

An old issue of the glossy Guangzhou magazine *New Weekly* features a woodcut of Lu Xun on the cover. "Everything we want to curse today," the headline reads, "Lu Xun has already cursed." This is still as true as it is tragic. A civil war, a communist experiment, and tens of millions of deaths later, China is still haunted by the same ghosts. They whisper about the shameless enrichment of a corrupt elite, the abuse of power, and repression. Lu Xun's words about the "man-eating society" crop up time and again in blogs and comment pieces that have managed to slip through the cracks in the wall of censorship. Today China stands before the same unanswered existential questions: how should it deal with the world; how should it deal with its own citizens?

China is civilization, and civilization is China—that's what the Confucians taught. After all, China was the Middle Kingdom, the heart of the civilized world. The ruler of China brought order to the world and peace to his empire, and put the peoples in the farther reaches of the world under his spell with the sheer force of his charisma. The Confucians also taught that "the ruler is the ruler; the subject is the subject." When everyone knows his place, harmony will reign. The communists once burned Confucius. Now they're pulling his books out of the ashes—collections of Confucian maxims have become bestsellers—and erecting altars to him once again. Confucian schools are opening all over the country.

Lu Xun and his contemporaries hated the Confucians. Or rather, they hated what China's state thinkers and rulers had turned Confucius into over the course of more than two millennia: a thinker whose philosophy had been robbed of its humanity, leaving only a tight corset of ritual and hierarchy. The rage of these reformers was directed first and foremost at that spirit of subservience based in Confucian philosophy, and the despotism, disguised with empty words about "humanity," under which the Chinese people had been forced to their knees for more than two thousand years. Lu Xun attacked the "slave mentality" of his people, and in 1919 the pioneer of liberal thinking Hu Shi called for people to "tear down Confucius and Sons!" Nowhere would you find more radical haters of Confucius than in the ranks of the Communist Party, whose Great Helmsman Mao Zedong boasted that he had outdone even China's first emperor. Qin Shi Huang, he proudly declared, had only had 460 Confucians buried alive; Mao had a thousand times that number on his conscience.

The Party's rehabilitation of Confucius, then, was no small thing. It began under Xi Jinping's predecessors, with the sudden rise of the old Confucian concept of harmony, *hexie*: the new high-speed trains were all christened "Harmony." ("Welcome to Harmony!" a voice would breathe in both Chinese and English over the loudspeaker, as the Beijing–Shanghai train departed. When it reached its final destination, the voice would bid you farewell with "Goodbye in Harmony!") The propaganda apparatus celebrated the ideal of the "harmonious society," meaning the harmony between order and obedience above all else. In 2011, to many people's surprise, an enormous bronze statue of Confucius appeared on the edge of Tiananmen Square, just a stone's throw from Mao's mausoleum. The creator of the reborn Confucius, incidentally, was Wu Weishan, who sculpted the Trier Karl Marx. The giant philosopher vanished just three months later without explanation: perhaps a sign that Confucius's return was still somewhat controversial.

It would not be until Xi Jinping became Party leader that the controversy would be put to bed. Not only did Xi pull off the feat of putting Mao and Marx back on the front row in the communist ancestor-worship shrine; he successfully brought Confucius to stand at their side. Xi was the first head of the CCP to follow in the footsteps of the old emperors and go on a pilgrimage to Qufu, the birthplace of Confucius. There he attended a celebration to mark the 2,565th birthday of China's most influential philosopher. In a speech to an audience of Confucius scholars, Xi explained why his Party's policies were the logical and inevitable results of the thought of old China. The ideology and culture of China today, he said, couldn't be seen as anything but "the continuation and refinement of traditional Chinese ideology and culture."[90] Confucianism, he added, is the key to understanding the "special characteristics" of China that the CCP so frequently invokes. And naturally, the Party officials themselves are the "inheritors and upholders of the country's fine cultural traditions."[91]

The speech had a certain chutzpah in light of the ecstatic orgies of destruction perpetrated by Mao's Party against anything with a whiff of culture or tradition about it. The same Party over which Xi Jinping now presides literally wiped out an entire generation of people who were upholding China's fine cultural traditions: artists and poets disappeared into labor camps, professors were forced to muck out pigsties, upper-class patrons and collectors were executed, and writers were driven to suicide. Meanwhile, the Red Guards burned books and demolished Confucian temples. The wonderful thing about Confucius, however—at least, according to Xi Jinping in Qufu—is how he has "changed over time and adapted to each new set of circumstances." In other words: how each ruler of China manages to gut the old philosopher to serve his own ends. Today, that might be easier than ever: the break with tradition and history forced on the country in the Party's first four decades of rule was so radical, and people's amnesia and ignorance are now so deep-rooted, their fear so great, that the weirdest and most absurd characters can be painted on that virginal sheet of paper with very little protest.

Xi Jinping has made it clear that he sees his rule as analogous to the mighty dynasties of old. His "China Dream" is the "great rejuvenation of the Chinese nation," restoring it to its former glory. At first it might seem surprising that the communist Xi Jinping, of all people, has taken to deploying nationalist reactionary rhetoric (a strategy not seen in China since Chiang Kai-shek) and invoking ancient Chinese thinkers, especially Confucius. The purpose of this doctrine, though, is to inject stability and morality into a directionless Chinese society. If everyone knows his place in the family, society, and the state, then order will assert itself.

Drawing on Confucius *et al.* is a tactic, intended to give the population back the identity and values it has lacked in recent years. It also helps Xi in his quest to promote the notion that China and its national circumstances are "unique," and thereby to immunize the country against the temptations of Western values and practices. Today universities up and down the country are not only turning with renewed vigor to the study of Marx and Lenin; they are simultaneously setting up large institutes for *guoxue* ("national learning"), which mine Chinese traditions for calendar mottos and other nuggets to aid the ruling dictatorship.

It is rare to hear voices like that of the historian Li Ling from Peking University, who accused academic colleagues of taking an approach that "religionizes, politicizes, commercializes, simplifies, and vulgarizes Chinese culture," with the sole aim of satisfying "some confused leaders."[92] In 2017, the CCP and the government published their own document on "implementing the project to spread and develop the great traditional culture of China."[93] Culture and tradition, it says, are "the blood vessels of a people and the home of its spirit." The unique ideas and the wisdom of Chinese culture give the Chinese people pride and self-assurance. They therefore make an essential contribution to building "a mighty socialist state with Chinese characteristics, to the cultural Soft Power of the state and to the realization of the China Dream of the great rebirth of the Chinese nation."

At the same time, in 2019, the party published "Guidelines for Building Morale for Citizens in the New Era," which prove instruction on "socialist core values" as well as an "ethics of cyberspace," the latter summarized as "disseminating positive content" and "monitoring social media platforms."

In any case, a leader does not convert to Confucianism without risk, because doing so also reminds people of the old Confucian ideal of the noble philosopher king: a ruler is so wise, virtuous, and just that his charisma alone inspires people and brings order to society. "He who rules through moral strength is like the pole star," it says in the *Analects of Confucius.* "He stays in his place and all the other stars circle about him." For the Confucians, a good king is no autocrat, pursuing power without limits; and a good government neither intervenes excessively in society, nor doles out punishments. Ministers and scholars have not only the right but the duty to contradict the ruler if he strays from the correct path. By these measures, Xi's rule is anything but Confucian. In this respect, too, Xi is like most of the old emperors: they all had Confucius on their lips, but their hands were firmly clasped on the dagger of draconian laws and punishments.

While we're on the subject of the past, let's not forget the incessantly repeated mantra of "5,000 years" of Chinese history. When I first went to study in China (less than 2,000 years ago) ceremonial speeches only ever mentioned 3,000 years. These 3,000 years are historically proven: the oldest Chinese dynasty for which we have archaeological evidence is the Shang Dynasty, probably originating in the 16th century BC. But at some point the CCP leadership must have heard about Egypt and its 5,000 years of continuous history—and Beijing doesn't like to come second in anything.

The propaganda machine tries to salve the pain of the West's technological superiority over the past hundred years with memories of China's glorious age, when the country invented paper, printing, gunpowder, and the compass. These are the "four great inventions," to which state-sponsored researchers are always adding further discoveries: in 2016 a team of more than 100 academics announced that it had discovered China was also responsible for the invention of the stirrup, the rocket, the decimal system, the crossbow—all in all, 88 scientific and technological breakthroughs. The icing on the cake was that the team took eight years to make their findings—by the happiest of coincidences, the Chinese regard eight as the luckiest number.

The continual discovery of historic Chinese inventions has been a popular sport among the Party press for many years. They have long since reclaimed football and golf for China, and announced that China also "invented Lassie," as one newspaper headline put it, following new information on the domestication of dogs 16,000 years ago.

Nationalism is a powerful tool. Wherever it takes root, as the British scholar Ernest Gellner once wrote, it easily retains the upper hand over all other modern ideologies.[94] Since it was founded in 1921, the Chinese Communist Party has also been a nationalist party; many of the early revolutionaries were attracted to communism as the force that would raise their country up again and make it strong. After the *de facto* death of communism in the 1980s, the Communist Party searched for new sources of legitimacy for their power and found two: one was the promise of material prosperity, and the other was nationalism. Patriotic pride in China's great cultural past—the poets of the Tang, the philosophers of the Song, the artists and authors of the Ming—was now no longer forbidden (as it had been under Mao). On the contrary, it was encouraged.

At the heart of the nationalist propaganda, however, there soon emerged a poisonous variant: the narrative in which the Chinese people are first and foremost victims. Schools, museums, speeches, films, and TV series present Chinese history as an endless series of humiliations by foreign powers, from which only the Communist Party has brought salvation. In this narrative, China is a nation with 5,000 years of great history, but the corruption of the Manchu dynasty (who were also foreign rulers) combined with the barbaric aggression of the Japanese and "the West" brought a "century of humiliation," starting with the Opium War instigated by the British in 1839. China was only saved from this age of indignities by the glorious victory of Mao's communists. After the Party had instructed the army to shoot its own citizens in the massacre of 1989, the need for fresh legitimation became even more pressing, and greater emphasis was placed on "patriotic education." Henceforth, China's children were to be injected with nationalism from a young age.

The victim mentality and the memory of humiliation by foreign powers are assiduously kept alive. A whole industry has sprung up around the hatred of Japan in particular. For a time it was impossible to switch on a TV without seeing Japanese soldiers committing murder, arson, and rape, before being dismembered, beheaded, or blown up by heroic resistance fighters. These series played on dozens of channels at once, at all hours of the day and night. In March 2013 the *Southern Weekend*, at the time still the country's bravest newspaper, counted no fewer than 48 TV crews all filming anti-Japanese TV series at the Hengdian World Studios.

One of the principal propaganda arguments of recent years is the one that says that the West is still out to sabotage China's rise, and it fits seamlessly on top of the old story. The message has been put out in numerous different ways—with a series of cartoons, for example, in which a sweet little rabbit (China) is persecuted by an American bald eagle. According to this narrative, any criticism of the Party or of the governance of China is purely an attempt to weaken the country. To make this argument stand up, the Party must claim perfect equivalence between itself and the nation: the Party *is* the people, it *is* China. Foreigners who flatter the Party may then be honored with the title "friends of China," the more critical "do not understand China," or "are hurting the feelings of the Chinese people." If you're Chinese and you love your country, it follows with an iron logic that you *must* love the Party. Anyone who criticizes it is a traitor.

Meanwhile, China has outwardly hung on to the slo-

THE DREAM • 221

gan of the country's "peaceful ascent" all these years. China needs the rest of the world for its economic miracle; and it needs stability and peace on its national borders. In the last two decades China has been one of the biggest beneficiaries of globalization, it is the country's aspiring middle classes who have profited most. This class may have swallowed and parroted the state-sponsored nationalism, but their growing prosperity helps to explain why so far they haven't turned against globalization. China's cities, and the millions of workers in the export factories on the coast, know well what they are gaining from cooperation with the world and an open economy.

Xi Jinping has added fuel to this nationalism. The "China Dream," *zhongguomeng*, which he made his personal trademark on taking office, had been the title of a nationalistic bestseller two years previously.[95] Its author, Liu Mingfu, a retired colonel in the People's Liberation Army, calls for the bolstering of the Chinese army and prophesies a growing rivalry with the USA, at the end of which China will be number one in the world. Ian Buruma, an expert on China and Japan, has written in the *New Yorker* that the Asian renaissance described in *China Dream* sounds "remarkably like the Japanese propaganda in the 1930s."[96]

The Party is playing with fire. The victim mentality

it has nurtured, and the idea of seeking refuge in the lap of the motherland, have become a reflex even in everyday life—as demonstrated in January 2018, when in the space of a single month, three Chinese tourist groups abroad attempted to summon the strong hand of their great nation. One party was stuck at the airport in Tehran, and had not been allocated the accommodation its members were expecting. In a video circulated online, you can see several hundred Chinese people venting their anger at this turn of events by chanting: "China! China!" Something similar had taken place a few days previously at Tokyo airport: after their flight home to Shanghai with a Japanese budget airline had been delayed by 24 hours, the tourists began a raucous rendition of the Chinese national anthem in protest as they tussled with Japanese police officers. A choir of frustrated Chinese passengers also belted out the national anthem at an airport in Sri Lanka.

This was embarrassing—and not just to the many ordinary Chinese people who posted comments on social media. It also made official China uneasy. The state broadcaster, CCTV, published a comment piece calling Chinese compatriots to order: "No one who displays an unseemly 'Wolf Warrior'-style patriotism and yells out 'China!' whenever he feels like it is going to earn any sympathy in a foreign airport, or from his countrymen at home. Just because the motherland is there to support you when you have problems does not mean that it will play the scapegoat for you."[97]

However, the Communist Party itself has long been trying, with some success, to use overseas Chinese for their cause, to make the Chinese diaspora in the West an instrument of its propaganda and a means of pressure against universities, civil society groups, or governments that it considers troublesome. The party has essentially kidnapped a broader pride in China and replaced it (alongside strict penalties for variance) with its version of patriotism, which is heavily dependent on Great Power nationalism and Han chauvinism. And it works, in that many are swept up in the narrative that China has been and still is the eternal victim of Western humiliation, and the Communist Party the only salvation, and any criticism of the CCP is nothing but an attempt to sabotage China's global ascent. It is remarkable to witness how powerful and effective such propaganda is when for most Chinese it has been a lifelong exercise starting in kindergarten. As a result, you see Chinese students on the streets of Hamburg and in the universities of Sydney, Cambridge, or Boston insult and threaten Hong Kong classmates, or in Toronto, where young Chinese wrapped their Ferraris and Porsches in Chinese flags as a sign of support for the Chinese

Communist Party, honking and vilifying Hong Kong demonstrators as "poor pussies." One probably can say goodbye to the illusion that students living in Europe or America for years would automatically succumb to the charisma of democracy and could turn out to be a force of change at home in their own country.

Within China, a noisy minority has adopted an aggressive form of nationalism, and its ugly face keeps appearing in the public sphere. It could be seen in Beijing in 1999, at demonstrations against the NATO bombardment of Belgrade's Chinese embassy. During the demonstrations, correspondents from CNN and the German channel ARD were physically assaulted. Or at the anti-Japan riots of summer 2012, when placards on the streets of the Chinese capital called for Tokyo to be wiped out and demonstrators in Xi'an beat up the Chinese drivers of Japanese cars.

You will find it online, too, where the *Xiao Fenhong* ("little pink ones") react to any hint of criticism of the motherland and the Party's patriotic myths by firebombing forums with ultra-nationalistic commentary. The *Xiao Fenhong* gather in places like the "emperor forum," which has almost 30 million registered members, and from there launch organized attacks on websites outside the Great Firewall. Thousands of these pink trolls suddenly appear on Instagram when Lady

Gaga meets the Dalai Lama; they flood Taiwan's news sites with obscene tirades against the Taiwanese president Tsai Ing-wen; and when students at the Chinese University of Hong Kong make a case for their city's autonomy or independence, the *Xiao Fenhong* take over the university website's comment pages, posting threats such as "Tonight the emperor forum will find you."

The "pink ones" are so called because they originated on the pink forums of the popular women's fiction website Jinjiang City. In 2008, their debates moved from literary to nationalistic matters: it was the time of the Olympic Games, which had been planned as a celebration of unadulterated pride in Party and motherland, and were being overshadowed by the violent unrest in Tibet's capital, Lhasa, and lengthy human rights debates in the Western media. At that time, internet forums saw an especially powerful surge of vituperative nationalism.

"Who sent you all that money from the motherland?" one person asked the ungrateful Tibetans: "Have you forgotten about all that?" Another applauded the violent way in which the demonstrations in Lhasa were put down. The government did a good thing, he said, by "cutting out this cancerous tumour." Someone else warned, "If you behave badly, then we'll take your

culture and stick it in a museum," while another com-
menter asked: "Why are we even talking? Separatist
scum should be killed." Meanwhile, a blogger lamented
the fact that foreigners had all been "brainwashed." It
was a reminder that the devastation wreaked by the
communists takes many forms and will no doubt be
borne by future generations. The poisonous seeds being
sown by the Party today will still be sprouting decades
from now.

The Party knows how dangerous this angry nationalism is; that it might one day turn against them. The
propaganda machine advocates a "rational patriotism,"
yet time and again Xi Jinping adds rhetorical fuel to
the flame of national feeling, invoking China's "sacred
territory," of which not "a single inch" will be surren-
dered.

China's nationalism very often revolves around three
specific issues: first, Taiwan, the island that has to all
intents and purposes been a separate state since the
end of the Chinese civil war in 1949, although the CCP
leadership makes its return to the Chinese motherland
seem more urgent with every year that passes. Taiwan
is the subject that excites emotion most easily on the
mainland—and at the same time, it's the problem least
likely to be solved. Taiwan has been a vibrant democracy for more than three decades, and its 23 million

citizens don't show the slightest inclination to submit to the yoke of the CCP. The nationalists' second holy grail is the chain of islands in the East China Sea which the Japanese call Senkaku, and the Chinese know as Diayutai. It was these islands that sparked the anti-Japanese riots in the summer of 2012. Third, China has staked a claim to the South China Sea in its totality, basing its claim on dubious historical arguments dismissed by its neighbors as nonsense, while in China itself they have attained the status of sacred truths.

Xi is the first Party leader in decades who doesn't stop at words. Xi acts. He might make trite speeches in which he declares the extraordinary finding that Chinese blood contains "no gene for hegemony or invasions" (the Xinhua News Agency calls this the Chinese people's "Historical gene of peace-loving").[98] But he is pouring vast sums into remodeling the army—it must be able to "win wars"—and, above all, into strengthening the navy. It is the Chinese navy that is busily creating a *fait accompli* in the South China Sea: building artificial islands in disputed locations, enlarging coral reefs into navy bases, intimidating the fishermen and coastguards of neighboring countries. In July 2016, when the International Court of Justice in the Hague declared China's territorial claims in the South China Sea to be largely void, China declared the judgment itself "null and void"—and simply carried on reclaiming land and militarizing the waters.

This escalating aggression in word and deed is not without danger for China's leadership and for the world. Xi Jinping is fueling expectations, not all of which he will be able to meet. Nationalism can easily grow into a monster that even its creator can't fully control. What if Xi's own rhetoric maneuvers him into a corner during some future conflict, preventing him from making any kind of compromise that could be seen as weakness at home? What would happen if, after nearly four decades of growth, China's economy found itself in crisis? What if one of the Party's legitimating pillars—endless economic growth and increasing prosperity—threatened to crumble, leaving only the other: nationalism? How would the CCP, and how would Xi Jinping, distract the people from this state of affairs? Today, you can already meet people in China (especially among the younger generation) who believe that war with the USA is inevitable.

Some acquaintances from Beijing told me a story about their eleven-year-old son. A few days previously they had announced with great excitement that the family would be able to afford a foreign holiday that year, and that he could celebrate his twelfth birthday in

Europe. "No!" the son cried. "I'm not going abroad! I will only celebrate my birthday in the motherland!" The father's bafflement was still evident on his face as he related the incident. "It's madness," he said. "What are they doing to my little boy?"

The Eye

How the Party Is Updating Its Rule with Artificial Intelligence

"The perfection of power should tend to render its actual exercise unnecessary."

Michel Foucault

I n University College London, on a chair behind a pane of glass, sits the skeleton of the philosopher and social theorist Jeremy Bentham, dressed in a shirt and frock coat. Before his death on June 6, 1832, Bentham had left his body to the university, instructing that his mortal remains be put on public display. His friend Thomas Southwood Smith performed the dissection of the body during a public lecture. But the mummification of the head, using techniques learned from observing Maoris in New Zealand, didn't quite go to plan. It was soon decided to spare the public the sight of it, and a more appealing wax head was placed atop Bentham's skeleton.

The students hurrying by cast fleeting glances at the philosopher. A few take selfies; many ignore him. For a while, you could watch all this live: Bentham was not merely stared at; he stared back. A webcam positioned above his skeleton picked up curious admirers and people rushing past, and streamed the images live on Twitter. The team behind the PanoptiCam Project, which was set up to research surveillance algorithms, among other things, called it: "Watching you watching Bentham." Jeremy Bentham was the founder of utilitarianism, as well as one of the leading exponents of the British Enlightenment, an advocate of rationality and freedom, and a pioneer of democracy and the liberal society. He was also the inventor of the panopticon prison, which to this day remains one of the most illuminating metaphors for the surveillance state. This state achieves perfection when it can stand by and watch as its subjects take over the job of surveilling themselves.

The invisible authority, which sometimes praises and sometimes punishes, which has the power to crush a person yet which shows how generous it is by not crushing them: organized religion has been working with this concept for thousands of years. You can't see it, but it can see you. At every moment, day and night. Which means that it might not be looking at you right now, but that doesn't matter. You only need to know that it could fix its gaze on you at any time, and you will begin to monitor yourself. Bentham's panopticon was an architectural attempt to bring this form of surveillance to the world of the Enlightenment, to perfect and rationalize it. He himself thought it both efficient and humane, and spoke of "a new mode of obtaining power of mind over mind, in a quantity hitherto without example."[99] He could imagine his panopticon being used as a school, a factory, or a hospital, but above all it seemed to make the perfect prison. Bentham called it "a mill for grinding rogues honest."[100]

The panopticon is both simple and brilliant: a ring-shaped building with many floors, all opening inward onto a central courtyard. Each of these floors is made up of a continuous row of small cells. The cells each have two windows: one looking out, to let in light, and one looking inward, into the courtyard, at the center of which is a slender watchtower. This tower has windows facing out in all directions, placed so that the watchman can see into every cell in the large outer ring, but can never be seen himself. Nor can the inmates of the cells see each other. "All that is needed, then, is to place a supervisor in a central tower and to shut up in each cell a madman, a patient, a condemned man, a worker, or a schoolboy. By the effect of back-lighting, one can observe from the tower, standing out precisely against the light, the small captive shadows in the cells of the periphery," writes Michel Foucault in his book *Discipline and Punish*.[101] "They are like so many cages, so many small theaters, in which each actor is alone."

Bentham died in 1832, and the panopticon in the form that he envisaged it was never built. Yet it was such a powerful idea that the philosopher Michel Foucault used it as a metaphor for the constrictions of modern society. If an individual is completely isolated, but most important, completely visible, says Foucault, then power functions automatically. "[T]he surveillance is permanent in its effects, even if it is discontinuous in its action; [. . .] the perfection of power should tend to render its actual exercise unnecessary." The watchman, in other words, doesn't even have to be there. The inmates of the cells merely have to believe he might be present: "So it is not necessary to use force to constrain the convict to good behavior, the madman to calm, the worker to work, the schoolboy to application, the patient to the observation of the regulations. Bentham was surprised that panoptic institutions could be so light: there were no more bars, no more chains, no more heavy locks."

All you have to do is isolate individuals effectively, and guarantee their visibility. According to Foucault: "He who is subjected to a field of visibility, and who knows it, assumes responsibility for the constraints of power; he makes them play spontaneously upon himself; he inscribes in himself the power relation in which he simultaneously plays both roles; he becomes the principle of his own subjection." Those who are naked before this eye take over surveillance of themselves.

Foucault wrote *Discipline and Punish* in 1975. At that point, no one could have guessed that technological progress would one day allow for the total surveillance of every single subject. One of the projects through which China wants to link up surveillance cameras and databases nationwide is called *Xueliang*. It is a play on the word for "sharp"—as in the sharp eyes of the masses, a slogan from the Mao era, when the entire population was spying on each other. The all-seeing eye, observing a population of 1.4 billion people from a watchtower looming above and seeing into everything, has finally become a real possibility. China's security apparatus is thrilled.

March 15, 2016, will become one of those dates that the human race will always remember. On that day, in a ballroom in the Seoul Four Seasons Hotel, machines defeated man in a way no one had expected. It was a competition not unlike that between IBM's chess computer Deep Blue and the then world champion Garry Kasparov in 1997, when the computer won. Only this time, the stakes were much higher. The competitors were playing Go.

Go is one of the world's oldest boardgames, invented more than two and a half thousand years ago in ancient China, where it was called *weiqi*. For a long time, gentleman scholars regarded it as one of the four arts that every cultivated man should master—alongside calligraphy, painting, and playing the Chinese zither. Go, which is played on a board of 19 × 19 intersecting lines, is infinitely more complex than chess. Even world-class players often have to rely on intuition rather than linear thinking. After the first two moves in a game of chess, ther are over 400 possible permutations; in Go, that number is around 130,000.

The Chinese gentleman-scholar Shen Kuo, a famous astronomer, mathematician, geologist, pharmacologist, cartographer, hydraulic engineer, and general of the 11th century, estimated the number of possible positions in a game of Go at 10^{171}. The figure, cited in his *Dream Pool Essays*, is greater than the number of atoms in the universe. On that March 15, South Korean Lee Sedol, the strongest player of the past decade, sat down at the board to play against AlphaGo, the machine developed by the London DeepMind laboratory, now owned by Google. It was the fifth and last match of the tournament, and it was the final blow. AlphaGo beat Lee Sedol 4–1.

To produce AlphaGo's algorithm, its creators had used two artificial neural networks that would train each other. To start with, they fed in 30 million moves by skilled Go players; then they made slightly different versions of the machine play millions of games against each other—and from these games AlphaGo learned entirely new strategies unknown to human players. AlphaGo's triumph in Seoul surprised many and shocked some; few of those watching had expected such an outcome for at least another decade. The victory triggered remarkably emotional reactions as, all over the world, suspicion hardened into certainty (even among laymen): the future is here. The technologies at the heart of AlphaGo are going to transform our world to a degree we haven't seen since the industrial revolution two hundred years ago.

Artificial intelligence is "the new electricity."[102] These are the words of Andrew Ng, the one-time head of AI research at Google and Baidu, who now teaches at Stanford. AI is not just one of many new technologies—it's *the* force used by algorithms working with all kinds of digital data, and in the future it will power every branch of industry as well as many aspects of our private lives. It will help reinvent everything that can be captured or supported digitally, whether in education, medicine, the world of finance, scientific research, transportation, or the city we live in. It is also reinventing the authoritarian state.

Nowhere in the world was the fascination and the surprise at AlphaGo's triumph as great as it was in China, the game's original home. And nowhere else reacted to it with such resolve and speed. China's leaders must have felt like the USA did on October 4, 1957, when it was caught off guard by the news that its great rival, the Soviet Union, had just launched the first man-made satellite into space: Sputnik 1. "Control of space means control of the world," Lyndon B. Johnson warned in 1958, when he was the majority leader of the US Senate. It was one of the key speeches of the Cold War. The future, he said, was "not as far off as we thought," and whoever won the space race would then have "total control, over the earth, for purposes of tyranny or for the service of freedom." Two AI advisors to the Chinese State Council described AlphaGo's victory in Seoul as China's very own "Sputnik moment."[103] And the Chinese state reacted in much the same way as the USA had in 1957: with a sudden change of tack, and a financial and strategic effort that almost overnight led to the funneling of huge resources into artificial intelligence.

"AlphaGo's victory fundamentally changed our thinking," says Zhang Bo, a well-known mathematician and AI expert from the Chinese Academy of Sciences. I saw Zhang Bo speak at an AI seminar in the modern conference center in Wuzhen, during the World Internet Conference at the end of 2017. Earlier that year, in this same city, in the southern Chinese province of Zhejiang, the DeepMind machine AlphaGo had given a repeat performance of its feat before a Chinese audience, beating the 20-year-old Chinese player Ke Jie (then at the top of the Go world ranking table) by three games to nil. After the match, the defeated Ke Jie, who is a celebrity in China, looked somewhat baffled. He later described AlphaGo as the "Go God."

The undisputed star of the conference that year was not the Apple boss Tim Cook, who was in attendance; nor was it China's chief ideologue Wang Huning, who brought greetings from his party leader Xi Jinping and announced that the digital economy was going to become the driving force behind China's economy as a whole: "We will not let this opportunity pass us by." The star was not even the internet itself—it was the miracle of artificial intelligence. "We will build a strong China with big data and artificial intelligence," said Wang Huning.

The euphoria of their Chinese hosts was echoed by the foreigners present in Wuzhen. The American IT expert John E. Hopcroft could be heard prophesying a third great revolution for humanity: there was the Neolithic revolution ten thousand years ago, when humans first became settlers; then came the industrial revolution; now the information revolution was upon us. Hopcroft had two main messages to give his rapt audience. First: thanks to AI, in the future only a quarter of the population would be required to produce and provide the sum total of all goods and services. And second, Hopcroft said, only two nations were going to profit from the AI revolution: the USA and China. Everywhere else lacked the infrastructure: when it came to AI, the threshold for entry was simply too high.

Delegates also heard Vaughan Smith, the Facebook vice president responsible for AI, introducing his platform (banned in China) to the audience. Smith was eager to assure his listeners, many of whom would be unfamiliar with Facebook, that, like all other technology, artificial intelligence would bring good things—and that in the Facebook laboratories, at least, they were "having a blast."

He underpinned his optimism with examples from Facebook's own AI research. Currently, they excelled at getting information from machines to people; the new challenge was how to transfer information from human brains to the machines. A brain produces one terabyte of data every second, the equivalent of 40 to 50 feature films shot in HD, said Smith. The big question is—how do you stream that data from living brains into whirring hard drives? At least, that is the question at Facebook, where according to Smith they are working day and night to find an answer.

If I understood Smith correctly, Facebook is doing all this with the noble aim of creating a world in which a person suffering from complete paralysis could operate a keyboard using just their mind and write 100 words a minute. Unfortunately, said Smith, the man on the street often had entirely the wrong idea about AI research, which was one of the reasons that the general tone of the debate was so anxious. He went on to report proudly on the project's latest coup: a cap that uses lasers to read the neurons in the brain.

A cap that can read your mind in real time, and siphon off all the films that play there into a machine? Isn't this just the kind of Bond-villain-style device the man on the street imagines AI scientists to be inventing—and is understandably very anxious about? It didn't seem to have occurred to Smith.

The Chinese members of the audience listened intently. "We extend a warm welcome to the international companies and institutes here, and invite them to open offices here in China and share the results of their research with us," said Chen Zhaoxiong, deputy minister in the Ministry for Industry and Information Technology (MIIT), with disarming openness. Afterward, representatives of Chinese industry presented their first AI successes. Lu Yimin, the president of China Unicom, one of the country's two telecoms giants, said that the age of isolated islands of data was over: "In the future, we'll have a central big-data platform." Having all the available information about each customer in one place—that was the way to follow through on the firm's new motto: "Be a customer-friendly creator of smart living."

"Smart" was also the key word for Robin Li, CEO of the search-engine company Baidu. In 2016, Baidu was hit by a scandal when it emerged that search results were being manipulated to the benefit of suspect advertising customers. Cancer patients in search of medical help were directed toward the website of a dubious group of clinics, and the death of a 21-year-old student with cancer triggered a debate about Baidu's practices that went on for months. With state support, the company is now trying to reinvent itself as an AI business, taking Google as its role model. Cars, supermarkets, cities—all will be smart in the future. "No more traffic jams, and no environmental damage. In the future, everyone will be relaxed and cheerful. We're going to make people happier," Robin Li enthused. He closed by saying: "We need to inject artificial intelligence into every corner of human life."

And that is just what China is doing. Xi Jinping has called for his country to become the "world leader" in AI as quickly as possible. Scientists must venture bravely out "into no man's land," so that China can also "occupy the commanding heights" in the area of artificial intelligence.[104] Barely a year after AlphaGo beat the South Korean Lee, and only two months after it beat China's Ke Jie, China's State Council published a "Next Generation Artificial Intelligence Development Plan."[105] It is an extraordinarily ambitious plan. Artificial intelligence, the authors write, will change human life and the face of the earth. AI has become "a new focus of international competition. AI is a strategic technology that will lead in the future." AI will turn the commercial world on its head and become the new engine of economic development. And finally, AI holds unprecedented "new opportunities for social construction," which is Party-speak for social control. An article in the *Guangming Daily*, the Party newspaper for intellectuals, described China's power and wealth at the time when the peoples of the world still lived in agrarian societies. "But then our country missed out on the industrial revolution" and fell behind the West. China is not about to make the same mistake with big data and AI: "Digitalization has given the Chinese people the opportunity of the millennium."[106]

Various players in China, from private high-tech companies and universities to cities, provincial governments, ministries, all the way to the military, had for some time been working on their own programs in the fields of big data and AI. But the State Council's plan hit them like a thunderbolt, and their efforts are now being combined and multiplied. By 2020, China aims to draw level with the "leading research nations," meaning of course the USA. In 2025, China wants to achieve its own "important breakthroughs" in AI research and use. And by 2030, Beijing wants China to be the only frontrunner and "the most important center for AI innovation in the world." At that point, according to the plan, the country's AI industry should be worth the equivalent of $131 billion. As an added bonus, "public security" will be more seamless than ever before, thanks to "intelligent surveillance, early-warning, and control systems." The plan calls for new "intelligent applications" for the "management of society," with the specific example of "video image analysis and identification technologies, biometric identification technologies, intelligent security, and policing products." As always in China, developments are coming thick and fast in the breathless rush to catch up and overtake.

The eye of the dragonfly is made up of 28,000 facets, each of them a little eye in itself. Dragonflies have a 360-degree view of the world and can pick up images five to six times more quickly than humans. *Dragonfly Eyes* is the title of a feature film that I watched on an autumn day in 2017, in an artist's studio in one of the faceless new estates north of Beijing's fourth ring road. "I'll tell you a secret," says a young man in the film. "I often watch you, on the monitor." Then, from off-screen, the narrator's voice: "This is a man," he says. "He will be seen 300 times each day. This is a woman. Her privacy is all used up. The man and the woman meet." It's a love story, to begin with. The woman works on a dairy farm. She watches the cows. And is watched by cameras. The people who monitor the cameras are also being watched, watching. The film shows a couple having sex in a car, it shows sweet nothings being whispered in a restaurant, a car chase on the motorway, wild punch-ups, and plastic surgery. The film has been screened at festivals, including the Toronto International Film Festival. It may be the first feature film for which not a single scene has been acted or specially filmed. Every image, every scene—over 600 of them—comes from Chinese surveillance and live-stream cameras.

China is eagerly bringing the future into the present, and at a much faster pace than the West dares to adopt. "In 2013, when I had the idea for the film, there was hardly any material," says the Beijing artist Xu Bing, who had become a global name in the late 1980s with his installation "A Book from the Sky," for which he invented thousands of new Chinese characters. "But then, in around 2015, suddenly all these streams started appearing online, on websites that anyone could access. A lot of them were surveillance cameras belonging to individuals and private companies. Suddenly we had much, much more material than we ever could have hoped for. It exploded at a rate no one could imagine." Xu Bing's team set up 20 computers to download images 24 hours a day. They took 11,000 hours of video and distilled it down to 81 minutes for *Dragonfly Eyes*.

A few kilometers from Xu Bing's studio is a functional room with screens on the walls—a lot of screens, with faces on them. They are our faces: from the street, from the corridor, each with a name, sex, and ID number. They have been captured by the cameras of Megvii Face++, one of the hottest start-ups in a hot sector, which claims it wants to change the world. Here, artificial intelligence is a business model. Megvii Face++ is already doing something the State Council plan has only just called for. In this room stands a young man named Xie Yinan. He's a marketing director. Xie Yinan is wearing sneakers, a T-shirt, cool glasses; he laughs often and speaks with the missionary zeal of someone who looks the future in the eye every day. "It's like being in a movie," he says. "I've been here three years now, and when I started, I couldn't have dreamed we'd be able to do the stuff we're doing today. All the things you've seen in science fiction films—we're going to make them a reality." He looks exhausted; it's been a long time since he had a day off. But he's also euphoric. His country wants to become the world number one in the field of artificial intelligence. And his company wants to become the number one in its field. These are the times we're living in. "We want to give the city eyes," says Xie Yinan. "And we want to make them intelligent."

Megvii and SenseTime, probably its biggest competitor, are based not far from each other in modern office blocks in the Haidian district of Beijing, cheek by jowl with the country's elite universities. The glass door at the entrance to SenseTime opens when the cameras recognize the face of the person approaching. One captures my face and projects it onto a screen in the reception area. "Male," it says next to my picture. "Age: 45." It's seven years out. "But you look so much younger," says the lady on reception. "Attractiveness: 99 percent," the screen caption adds. It occurs to me that the algorithm might just have a flattery mode for potential customers programmed into it. Then the screen shows my mood: "Annoyed." Well, overtired might be more accurate. And then a video screen starts to play an ad that the system has selected for me: Wuliangye, a millet-based *baijiu*, 52 percent proof. Just what grumpy 45-year-olds with money and power would drink in China.

Thanks to the technology of SenseTime and Megvii, people were able to unlock their Huawei or Vivo smartphones using their faces long before Apple came up with the idea. They can also use their faces to order fries in KFC in Hangzhou. Or pay for their shopping using the Alipay app, as more than 800 million Chinese citizens already do on a regular basis. Hotels in China use Megvii cameras to check that guests really are who they say they are. Train stations in cities like Guangzhou or Wuhan only allow entry to people once their faces have been scanned and checked against the police database. The company is currently testing unmanned supermarkets. "Thanks to our cameras, we can tell how old customers are, whether they're physically fit and what brands they wear," says Xie. "And judging from the things they buy, we can categorize them as a certain type of person, and then target them with specific ads and special offers."

Another group has been taken through the exhibition room ahead of us, one of whom is wearing a police uniform. The guide points to a screen on which a great crowd of people is moving; it looks like the crowd is swaying back and forth, and the people on the screen are highlighted in various shades of red and green. It looks like an infrared picture. Among other things, the guide tells us, the system can predict the movements of crowds. "We've sold it into a lot of provinces, and it's being used in Xinjiang, too . . . We've had some good feedback." Xinjiang is the troubled province in western China that is home to the Muslim Uighurs.

The eyes of the city. The eyes of the Party. For the individual, his or her face becomes a key, opening the door to the world outside. For the observer, the camera becomes a key that unlocks the world inside the individual, and their behavior. "Criminals today need to think hard about whether they're going to keep committing crimes," says Xie Yinan. "Our algorithm can support networks of 50,000 to 100,000 surveillance cameras. We can tell you what kind of person you'll find in a given place at a given time. We can ask: 'Who is that? Where is he? How long is he there for? Where's he going now?' We track a person from one camera to the next." The system, says Xie, is already much better at recognizing faces than people are. The new cameras also increase accuracy by combining facial and gait recognition, as each of us has a very distinct walking style.

Megvii and the other companies advertise their commercial applications, showcasing apps that magic a funny dog's nose onto your face. But the identity of their most important investor and their biggest customer is no secret: it's the state, and in particular the security services. Xie talks about receiving letters of gratitude from police stations all over the country. Some 3,000 criminals, whose faces were stored in the authorities' databases, have fallen into the laps of the police over the course of the last year, he says, thanks to the cameras. The 2017 Qingdao International Beer Festival made headlines: 25 people who had spent a long time on wanted lists were arrested as a result of facial recognition. In April 2018, in a stadium in Nanchang, cameras picked out a 31-year-old whose profile had been placed on a national database for "economic offenses"—from a crowd of 60,000 concert-goers.

The cameras can do more: they report when a face turns up at a particular place—a bus stop, for instance—with suspicious frequency. "That could be a pickpocket," says Xie. At SenseTime, a few blocks away, they also demonstrate how the cameras analyze crowds. The system can tell when a lot of people are gathering, says the company's spokeswoman Yuan Wei. And when a lot of people are *about to gather.* The algorithm can also see when a lot of people are moving in one direction, while a single individual is going against the flow. "The system then identifies this person as abnormal," says Yuan Wei. And it sounds the alarm.

In 2017, Megvii had around 200 employees. Two years later, there were more than 2,000, many of whom have returned home from the USA. China isn't just trying to import AI technologies and to buy out firms in the USA and Europe; it is also actively head-hunting AI talent, in growing competition with Silicon Valley. The state plays a central role, having launched a "thousand talents" program, which provides attractive incentives and benefits to those willing to settle and work in China. In late 2017, the Ministry for Science and Technology launched a project for "transformative technologies," among whose aims is to develop new high-performance chips by 2021, in order to power neuronal networks.

Unlike earlier large-scale science projects—in biotechnology, for instance—Beijing knows that the field of artificial intelligence is too broad, too diverse, and too dynamic to be driven forward effectively by brute force from the state, with planning bureaucrats issuing and pushing through top-down edicts. The contribution of private high-tech companies and countless start-ups, and their cooperation with the state, are central to Beijing's plans. The government is keeping these companies close: the Ministry of Technology has officially selected firms like Baidu, Alibaba, Tencent, and the speech-recognition company iFlytek to lead the development of nationwide AI platforms in areas such as self-driving cars, smart cities, medical diagnostics, and speech recognition. This gives the chosen firms an advantage in these markets, with valuable access to state databases. At the end of 2018 the Chinese Academy for Information Technology, a think-tank operated by the Ministry of Industry and Information Technology (MIIT), produced a "white paper on AI security" praising the large private internet companies Alibaba, Tencent, Baidu, and Netease for their active contributions to the "intelligentization of national social governance," in fields such as "security monitoring, data investigation, and public opinion control."[107]

The private sector, generously supported by state funds, has begun to open its own laboratories in Silicon Valley and elsewhere, allowing Chinese companies to woo foreign AI experts with the promise of great salaries and even greater opportunities. Come to China; you can do more there. And you can do it faster. The West is getting tangled up in legal restrictions and data protection concerns, while China just goes ahead with things. In November 2017, Megvii gathered $460 million in a single round of investment—at the time busting the world record for an AI start-up. Since then, the firm's technology has won several competitions, beating teams from Google, Microsoft, and Facebook. Since 2015, SenseTime's research team has presented more new studies at the world's major AI conferences than Google or Facebook, and in April 2018 it overtook its competitor Megvii when it brought in $600 million of new investment, becoming the highest-valued AI start-up in the world. Similar start-ups are springing up like mushrooms all over China. Much of their seed funding comes from state investment sources. And the money arrives quickly: in China, it takes little more than nine months for a newly founded start-up to see the first payment from the investment funds; in the USA, the average is currently just over 15 months.

"This is just the beginning," says Xie Yinan. "The market is growing rapidly. Competition doesn't worry us, the demand is huge, there's room for everyone." In 2016 there were 176 million surveillance cameras in China. At that point, the USA had 62 million— more per head of population than China. But here too, the ambitious nature of Chinese plans and the speed at which they're implemented are making China the frontrunner: by 2020 those 176 million cameras will have become more than 600 million, many of them equipped with AI technology. Very soon, SenseTime is planning to invest in five supercomputers, to run the "Viper" system it has developed. Viper will apparently be able to automatically monitor and analyze the data from networks of up to 100,000 cameras.

The Communist Party has discovered a new magic weapon in big data and artificial intelligence—that much was evident from Party and state leader Xi Jinping's New Year's address to the Chinese people in 2018. As he does every year, Xi sat in front of a large wall of books; the observant viewer could make out Homer's Odyssey and Hemingway's The Old Man and the Sea on the shelves. And as they do every year, internet users subsequently took a magnifying glass to each book spine.

Many noticed that this year, Xi Jinping had brought *The Communist Manifesto* and *Das Kapital* down within reach. Most significantly, though, for the first time two bestsellers on artificial intelligence had been given a prominent position on the shelves: Brett King's *Augmented* was sitting next to *The Master Algorithm* by Pedro Domingos. Domingos researches AI at the University of Washington, and his book invites people to join him in the search for the algorithm to end all algorithms, an algorithm that will go on developing itself infinitely. "If it exists," Domingos writes, "the Master Algorithm can derive all knowledge in the world—past, present, and future—from data. Inventing it would be one of the greatest advances in the history of science."[108] Of course, there is no guarantee that Xi has really read King and Domingos—unlike Marx.

In the USA, researchers and companies are following China's plans with curiosity; some are sounding the alarm. "They will have caught up with us by 2025," prophesies Eric Schmidt, who is now the head of Google's parent company Alphabet and has evidently studied the State Council document from Beijing. "And by 2030 they'll dominate the industry." Now, one should not take at face value everything that comes out of the mouths of Chinese planners, but in the past the Chinese state has shown what it's capable of when something has been identified as politically desirable. Take, for example, the rapid creation of the largest high-speed train network in the world. And the CCP has clearly fixed on artificial intelligence as the key to its own survival and the perfection of its rule.

In the field of technology, at least, China is disproving those skeptics who still think an authoritarian system is poison for any kind of innovation. "There's this strange belief that you can't build a mobile app if you don't know the truth about what happened in Tiananmen Square," Kaiser Kuo once said. Kuo is an American China-watcher and heavy metal guitarist, who spent many years as the head of international PR for Baidu in Beijing. "Trouble is, it's not true."[109] For a long time now, China has been much more than a paradise for fake Nikes and alleged Louis Vuitton handbags with Chanel logos stuck on them. It's also much more than just the "world's factory": the country with the cheapest labor, assembling our TVs and smartphones, the country that makes just $7 out of every $880 iPhone we buy.

After the USA, China has long been number two in the world when it comes to spending on research and development, and according to the latest report by the American National Science Foundation, those positions will soon be reversed. Between 2000 and 2015, China increased its research and development spending by 18 percent every year. Spending in the USA, by contrast, rose by an average of just 4 percent a year in the same period.[110] The report also says that since 2003, the number of scientific publications from China has quintupled. China's researchers have now outpaced the USA with their supercomputers, at least in terms of the number of such machines that appear in the world's top 500. In 2020, researchers in the harbor city of Tianjin are planning to bring the first exascale computer into service, which would be ten times as fast as today's supercomputers. (The USA hopes its Aurora project will have reached this point by 2021, a year later.) China is also on the heels of the United States when it comes to quantum computer research. And in October 2019, Xi Jinping personally chaired a study session of the Communist Party Politburo on blockchain technology, where he stated that China had to set international standards for blockchain "to increase China's influence and rule-making power in the global arena."[111]

Behind the Great Firewall—shielded by the censors not just from the free flow of information, but helpfully also from the competition—a high-tech parallel universe has come into being. China is the world's biggest market for e-commerce, and at least in terms of market value, Alibaba and Tencent have long since moved into the same league as Amazon, Alphabet, and Facebook.

In the fields of electronic transactions and the development of smartphone finance apps, China is way ahead of the rest of the world. "You get the sense when you leave China these days that you are going backward," the *Financial Times* observed. The economic opportunities and the technological leap forward are one thing. But just as enticing for the state is an unprecedented opportunity to look inside its citizens' heads, and to monitor them in their beds, on the streets, on their shopping trips, and during the most intimate of conversations.

WeChat, for instance, is a phenomenon that doesn't exist in the West. The app began as a messaging service; the Chinese WhatsApp. But for China's mobile phone users, it very quickly also became the Chinese Facebook. Then the Chinese Uber and the Chinese Booking.com and the Chinese Deliveroo. On smartphones at least, WeChat has to all intents and purposes gobbled up the Chinese internet. You can use WeChat to talk to friends, book taxis and hotel rooms, order food, buy movie tickets, book trains and flights, rent city bikes, choose a cable TV package, pay your water and electricity bills and parking tickets, and get fast credit. Most significantly of all, you can make cashless payments. And all the while, the state security services are looking over your shoulder. In 2019, more than a billion Chinese had WeChat installed on their phones.

Two apps, Tencent's WeChat and Alibaba's Alipay, have split the market for cashless payments between them—and the Chinese love it. An entire population has switched to mobile payment in record time. Hardly anyone uses debit or credit cards any longer—and hardly anyone carries cash. In 2017, the Chinese used their phones to make $17 billion worth of transactions. That year, over 60 percent of all cashless transactions worldwide took place in China. I could pay with WeChat in the snack bar in my side-street, the grocery store, the hairdresser, and use it to buy noodle soup for the equivalent of $1.50. Eventually, people started giving me funny looks when I reached into my trouser pocket for cash. In Beijing, even the beggars now use barcodes, which passers-by can scan using WeChat to make their small donation.

At a courthouse in the Haidan district of Beijing, you can use WeChat to submit files and pay fees. The identity of the person submitting is confirmed via facial recognition.[112] And in December 2017 the state press announced pilot projects in 26 cities to test WeChat as a state-recognized, electronic social-security identification and ID card.[113] It's the dream of every lazy citizen. It's also the dream of the surveillance state, which gets news of its citizens' every move and every transaction delivered for free in real time.

Tencent cooperates very closely with the censors, police, and state security. During a chat, you might suddenly realize that the conversation has stopped making sense: certain words are being automatically deleted by WeChat between sender and receiver, without either party being informed. In the city of Puyang, a construction supervisor named Chen Shouli spent five days in a cell after forwarding a joke on WeChat about a rumored affair between a singer and a senior government official, captioned with "haha." And Wang Jianfeng from Shandong went to prison for 22 months after calling Party leader Xi Jinping a *baozi* and a "Maoist thug" on WeChat. A *baozi* is a large steamed bun, and has been one of the portly Xi's nicknames for many years.

In 2017 China had more "unicorns"—start-ups with a market value of $1 billion or more—than the United States. Some experts believe that China's Huawei is ahead of Apple in the development of mobile AI chips. And, as we've seen, on January 17, 2018, a Chinese computer also managed to win against the Go master Ke Jie: Tencent's Fine Art software gave Ke Jie a two-stone head start, and still beat him.

But China's companies are still weak when it comes to foreign trade, and the country still trails behind the USA in terms of its talent pool and total investment in the high-tech area. The Dutch academic publisher Elsevier and Nikkei in Japan publish a list of the institutions whose AI research is most frequently cited worldwide: two of the top three are now from China, with three in the top ten. It pays to look beyond the numbers, though. According to Jeffrey Ding, researcher at the University of Oxford's "Governance of AI" program, and author of a study on Chinese progress in AI,[114] the nation's AI research is no longer lacking in quantity, but it may still lack originality.[115] His verdict: "China [. . .] still cannot match the leading countries in the most innovative research and the most talented researchers." Ding's 2018 study presented an "AI potential index" for the first time, to compare China's AI capacities with those of the USA—and the USA still performed twice as well.

The USA might still have the lead when it comes to researching the principles of AI, counters Xie Yinan from Megvii, "but in its practical application, we're already a long way ahead." He pauses, then: "The state doesn't place so many limits on us here . . . The government is behind us." Judges in Hebei Province are already getting AI to help them prepare their verdicts, and according to the state press, courts and lawyers in Shanghai are using it to check the quality of evidence, and avoid "convicting innocent people."[116]

The city of Hangzhou is using algorithms to predict traffic flow, and all over the country "smart city" projects are being launched in cooperation with firms like Alibaba and Huawei. The CCP, meanwhile, is going even further. In the province of Sichuan, and within its Youth League, the Party is testing a "smart red cloud" on itself: the Party press says the algorithm is designed to modernize the way it assesses and chooses its functionaries. The system tracks the behavior and "human relationships" of CCP cadres, in order to predict "their future ideas and future behavior." And according to the Clean Governance Centre at Peking University, big data is also set to become China's "most powerful weapon in the fight against corruption."

Some in the USA are seeing parallels with the space race against the Soviet Union, more than half a century ago. And many believe China has the advantage: for one thing, while China is focusing on clear strategic goals, the USA is foundering under the presidency of the ignorant, anti-science Donald Trump. (Trump presented his government's AI strategy in February 2019, 19 months after Beijing's leaders.) But most important is the sheer mass of data to which researchers and companies have access in this nation of billions, with little hindrance from laws or debates about privacy. If the oil of the future is data, as the *Economist* puts it, then China is the new Saudi Arabia. The camera manufacturer SenseTime claims to have access, via government departments, to databases containing 500 million faces. Its competitor Yitu even claims it can compare 1.5 billion faces. That includes not only every Chinese citizen, but also every foreigner who has passed through China's borders, and whose features are automatically saved to a central database from the snapshot taken by border officials. "No other country can compete" with China's advantages here, according to a paper produced by the Beijing technology investor Sinovation Ventures: "In China, people use their mobile phones to pay for goods 50 times more often than Americans. Food delivery volume in China is 10 times more than that of the United States. And shared bicycle usage is 300 times that of the US."[117]

On this last point, it is worth knowing that the candy-colored rental bikes you find everywhere in China send all their movement and payment data back to a central database. In 2018, the company Mobike announced that its eight million bikes provided 30 terabytes of data every day. The company not only knows when and where and how fast you are cycling; it also analyzes who is cycling with whom. And it shares all that data with the government.

The more data is collected, the more fodder the self-teaching algorithms have to perfect themselves. Not all experts agree, though, that the sheer mass of data makes other factors, like the quality of the AI semiconductor technology or the number of AI experts, unimportant. According to the Oxford University study mentioned above, in 2017 there were 78,700 scientists working in the field of artificial intelligence in the USA, while in China it was 39,200. The study doesn't rule out the possibility that, for the smarter AI algorithms of the future, access to a greater mass of existing data might be less important.

One advantage China definitely enjoys is the speed and the lack of restraint with which it can commit itself to developing this technology. "Artificial intelligence was invented in the West," says the *MIT Technology Review*, the magazine produced by the elite US university in Massachusetts, "but its future is currently taking shape on the other side of the world." In May 2018, SenseTime announced a partnership with MIT; SenseTime founder Tang Xiao'ou had earned his doctorate there in 1996. His company was only set up in 2014, and today it's already worth more than $7 billion. SenseTime has lured people from MIT, Microsoft, and Google to work for it in Beijing. "We're still a baby, compared to Facebook and Google," says the PR representative Yuan Wei, "but our aim is very clear: we want to become world leaders." The company's Chinese name is Shang Tang, after the first Chinese dynasty, Shang (18th to 11th century BC) and its first emperor, Tang. "That was a time when China led the world," Tang Xiao'ou once said, explaining the choice of name, "and it will again, with its technological innovations."

The Party foresees a wide range of uses for AI. Whether in education, public health, or infrastructure, the aim is for new technologies to solve problems and increase productivity. At the same time, it wants to use artificial intelligence to provide central planners with a feedback and steering mechanism, with which it can predict and prevent any potential economic and social crises that might threaten the system. "The widespread use of AI in education, medical care, pensions, environmental protection, urban operations, judicial services, and other fields will greatly improve the level of precision in public services, comprehensively enhancing the people's quality of life," says the first part of the state AI plan.[118] It continues: "AI technologies can accurately sense, forecast, and provide early warning of major situations for infrastructure facilities and social security operations; grasp group cognition and psychological changes in a timely manner; [. . .] which will significantly elevate the capability and level of social governance, playing an irreplaceable role in effectively maintaining social stability." AI promises to make the dream of all authoritarian rulers come true: control and surveillance of the entire population. Politburo member Chen Yixin, who is responsible for law enforcement agencies in the country, said in a speech to party officials that they had to "place the process of AI governance development in an even more important position, elevating it as an important means of control."[119]

Facial recognition has made the most headlines when it comes to control. Universities are using it to check student attendance, and dispensers in Beijing's Temple of Heaven are using it to ensure the thrifty use of toilet paper: the machine releases 60cm of paper per face. Anyone who requires more has to wait nine minutes before another 60cm are granted. The Shenzhen police advertised the fact that they had solved a case of child abduction within 15 hours thanks to facial-recognition cameras. In the same city, the metro trains have cameras installed in every car, which can monitor "every inch of the train in ultra-high definition," as the *South China Morning Post* reported. "Not only can passengers' every move be closely watched, but their most subtle facial expressions are being captured and transmitted in the form of ultra-clear images, without any delay whatsoever."[120]

The transport police in Shanghai are using intelligent cameras to catch people driving without a license; registry office clerks in Chongqing are exposing people who commit marriage fraud; and the police in Jinan and Shenzhen are publicly shaming people who cross the street when the lights are red. Their faces appear in real time on a video screen at the side of the road— together with their name, address, and ID number. At a single crossing in the district of Fujian, the cameras caught no fewer than 13,930 people in the space of ten months. Very soon, rule-breakers will start receiving automatic texts to their mobile phones, telling them off and warning them that next time, they are liable to be punished.

Middle School No. 11 in Hangzhou drew enthusiastic attention from the press when it had "eyes in the sky" installed in every classroom: surveillance cameras with a continuous view of every student.[121] "They are all-knowing eyes; nothing gets past them. As soon as someone nods off or starts daydreaming, he is captured on the spot, using facial recognition," said an article on Sina.com. According to the article, the cameras not only capture how often during the eight-hour school day a student's mind wanders; they also "analyze facial expression and mood—whether someone is happy, sad, annoyed, or reluctant—and send the data straight to a terminal that analyzes the student's attitude to learning. The system really does have magic powers." The school has long since done away with the school card that students used to use for the cafeteria or the library. "Students scan their faces to get food, they scan their faces to buy things, and they scan their faces to borrow books." Big data and facial recognition, according to the report, are helping "students to study more efficiently." A student named Xiao Qian admits that he used to be a bit lazy in the lessons he didn't enjoy as much: "You might close your eyes for a minute or read another school book under the desk." With the eyes in the sky, those days are gone: "Now you feel the gaze of a pair of mysterious eyes on you constantly, and no one dares to go off-task any longer."

Years ago, China's police force christened its nationwide network of cameras "Skynet"—quite probably without any hint of irony or reference to the *Terminator* films, where Skynet is a rogue AI organism on a mission to wipe out the human race. China's Skynet, the press reported, is "the eyes that watch over China." In March 2018, a tweet from the *People's Daily* claimed that now, the Skynet was "capable of identifying any one of China's 1.4 billion citizens within a second," and was thus helping "the police in prosecuting crimes." Masks, hats, sunglasses, even plastic surgery present no problem these days, claim Megvii and SenseTime. "Machines aren't as easy to fool as people are."

The USA's experience with this technology has been rather different. An investigation by the Government Accountability Office brought to light such serious deficiencies in the FBI's facial recognition program that in March 2017 there was a hearing on the matter in the House of Representatives. In 15 percent of all cases, the FBI's technology identified the wrong people when searching for criminals, with black women being the most likely group to be affected by these errors. Such false alarms have more serious consequences in the Chinese system than they might elsewhere, since the conviction rate is over 99.9 percent. The state apparatus is always right, and where evidence is lacking, the police often provide the court with a forced confession elicited with threats or torture.

We can assume the system is still a long way from being all-encompassing: databases that often don't communicate with each other; firms that overstate their achievements; technology that promises more than it can currently deliver. And yet all the overblown claims—that the cameras can identify someone at a distance of 15 kilometers, or that everyone can be identified within a second—still make sense. The all-seeing eye doesn't have to be looking at you for the panopticon to function. All that matters is that you feel it might be—even if in reality, it isn't there yet.

China's cities are already the most monitored in the world. In 2019, a study by the Comparitech web portal counted eight Chinese cities among the ten cities across the globe with the most surveillance cameras per capita.[122] (London came in sixth, Atlanta tenth.) Chongqing was in first place with 2.6 million cameras installed (168 per 1,000 inhabitants), and in second place was Shenzhen with 159 cameras per 1,000 inhabitants. However, the report highlighted the plans of the Shenzhen city government to increase the number of its nearly 1.9 million cameras to almost 17 million in the next few years—this with a population of just under 13 million people. After the outbreak of the coronavirus, two companies, SenseTime and Megvii, were quick to market thermal cameras that combine fever measurement of passers-by with face recognition. By the end of March 2020, the cameras were already ubiquitous in the subway stations, schools, and shopping malls of major cities. SenseTime claimed that its cameras could identify faces "with relatively high accuracy" even behind the face masks that were omnipresent at the time.

At the height of China's coronavirus crisis, Alibaba in Hangzhou was the first of the country's major internet companies to develop so-called Health Code apps for the authorities. Others followed and the apps were quickly deployed throughout the country. These apps give smartphone users a green, yellow, or red code indicating the state of health of the user. Guards at subways, shopping malls, restaurants, and hotels constantly checked the codes of passengers and clients: only people with a green code, those who were considered to be healthy and not infected, were allowed to move relatively freely. A yellow code in most places meant that the person should be at home in isolation. And people with a red code were considered potential or confirmed carriers of a virus infection and should be quarantined. In addition to Alibaba's Alipay version of the app, many Chinese people used the WeChat's version developed by Tencent. According to state media, in April 900 million people had registered with WeChat's Health Code app.

Some of the information in the app is provided by the users themselves (signs of coughing or fever). Additionally, the app uses information provided by the authorities (medical records) and travel history as recorded by the smartphone: who has been where and when and who has had contact with whom for how long. People who are on the move even though they have been classified as red are penalized with an entry in their social credit system file. The government of Heilongjiang Province, for example, is threatening "fraud, concealment, and other behaviors" with punishments that would have "a huge impact on their future life and work." Concerns around these apps were not only caused by reports that the apps are programmed to forward information directly to state authorities without the knowledge of the users, but also by the fact that they in many cases came to the wrong conclusions: some healthy Chinese ended up being assigned a red code, without any indication why or how to get rid of it. At the same time, however, a large part of the population participated voluntarily and eagerly. In this case, the lofty goal of public health gave a further boost to the normalization of high-tech surveillance by tracking tools—a development that we also witnessed in the West a few months later. However, unlike in China, in most European countries tracking by coronavirus health apps is at least passionately discussed and accompanied by legal restrictions.

In China, tracking tools and the Skynet are just part of a much more comprehensive "police cloud," on which the organization Human Rights Watch (HRW) has reported.[123] The police cloud is a project from the Ministry of Public Security, which issued instructions for its use in 2015. Since then police forces in various provinces have been gathering all the data they can on hundreds of millions of citizens: medical histories, takeout orders, courier deliveries, supermarket loyalty card numbers, methods of birth control, religious affiliations, online behavior, flights and train journeys, GPS movement coordinates, and biometric data, face, voice, fingerprints—plus the DNA of some forty million Chinese people, foremost among them the Uighurs in Xinjiang. The state press reported, for instance, that the police force in the city of Xuzhou in Jiangsu was buying information from private companies, including "navigation data on the internet, [and] the logistical, purchase and transaction records of major e-commerce companies,"[124] but also the MAC addresses of the devices through which individual internet users could be identified. "Officers used to go house to house to gather information," the article says. "Today, the machines collect it tirelessly throughout the day."[125] Since 2019, you cannot be issued a SIM card if your phone is without a registered facial scan.

"Big data shows us the future," wrote Wang Yong-qing (at the time general secretary of the powerful Party Central Political and Legal Affairs Commission) in 2015, in the Central Party School's paper *Qiushi*.[126] The Party had to "assemble a complete collection of basic information about all places, things, issues, and people, tracking trends in what they eat, how they live, where they travel, and what they consume." All this would make "our early-warning system more scientific, our defense and control more effective, and our strikes more precise." Wang's boss at the time, Politburo member Meng Jianzhu, called on the security services in autumn 2017 to remove all barriers to a broad exchange of data: as soon as possible, he said, the images from surveillance cameras everywhere in the country needed to flow into a single database.

Meng's exhortation alludes to the age-old problem faced by large bureaucracies, especially in authoritarian systems: each subordinate authority reproduces the secrecy and lack of transparency of the apparatus in miniature, partitioning itself off from the authorities alongside it. The recurring complaint about isolated "data islands" urgently needing to be joined up suggests that the head-office strategists had a tough job implementing their plans on the lower levels.

All the same: AI, said Li Meng, Deputy Minister for Science and Technology, in summer 2017, would help China "to know in advance who might be a terrorist, and who might be planning something bad." Prophesying future crimes used to be the stuff of science fiction novels and films, like Steven Spielberg's *Minority Report*. In China, they're trying to make it reality.

As early as 2016, the *Shandong Legal Daily* was reporting on a project being run by the police in the city of Dongying.[127] The project is called "Mornings at Eight" and is cited by the paper as an excellent example of how to use the police cloud creatively. Every morning at eight o'clock, the system sends a report to the mobile phones of the 1,300 participating police officers, detailing the "abnormalities and trends" seen in their area. The algorithm, so the paper says, analyzes the data from the previous day, including information from hotels, internet cafés, airlines—and from the police force itself. It scans all data on any newcomers to the district, for instance, including their hometown, ethnicity, previous convictions, and online behavior, harvested from internet cafés. "Every day at eight o'clock [. . .] the system sends us targeted messages," the paper quoted a police officer as saying. "In particular, it alerts us to individuals who are involved in terrorism and in [undermining] social stability who've entered our jurisdictions."

Itinerant laborers and ethnic minorities appear as people of interest in these alerts with particular frequency. The HRW report quotes a tender document from the Tianjin police, which says its police cloud can monitor "people of certain ethnicity," "people who have extreme thoughts," "petitioners who are extremely [persistent]," and "Uighurs from South Xinjiang." The system, according to this document, "says it can pinpoint the residences of these individuals and track their movements on maps."

The question is: what is a crime? And who may find themselves attracting unwelcome attention in a country where someone like the late author and Nobel Peace Prize laureate Liu Xiaobo is viewed by the authorities as a mere "convicted criminal"?

The aim of all the data-gathering by the Chinese police, according to a 2014 notice from the ministry of public security, is to create an "early warning" system that will alert them to any "abnormal behavior" by citizens. Sophie Richardson, the China director at Human Rights Watch, says the authorities "are collecting and centralizing ever more information about hundreds of millions of ordinary people, identifying persons who deviate from what they determine to be 'normal thought.'" Ministry documents single out groups to be identified and targeted: not just drug addicts and people with criminal records, but also "petitioners," "those with mental health problems who tend to cause disturbances," and "those who undermine stability." In other words, people who, for one reason or another, are a thorn in the Party's side. Or who might become one, even if they themselves don't know it yet. Especially them, in fact.

To make absolutely sure that nothing goes wrong, the government is getting directly involved with some of the new companies. The Chinese state is the majority shareholder in the Hangzhou firm Hikvision, which began life in a state research institute and is now the largest manufacturer of video surveillance systems in the world. Today, Hikvision cameras have been installed in more than 150 countries, where they identify faces and license plates, and send an automatic alert when a driver is seen texting at the wheel. In the UK, 1.3 million cameras from the company were in use in 2019. In the USA, prisons, airports, schools, private dwellings, and military institutions are equipped with the Chinese firm's systems. Even the US embassy in Kabul installed one.

Their use has caused much debate in both Washington and London. Critics point to the fact that Hikvision works closely with China's Ministry of Public Security and its research institutions, while the company has frequently tried to draw a veil over its relationship with the Chinese regime. Its managing director, Chen Zongnian, is also the firm's Party secretary, and in March

2018 he was a delegate to the National People's Congress for the first time. Some suspect that Hikvision may have equipped its technology with a "back door," which would allow China to access its internet-enabled cameras and channel their data to Chinese servers. So far, the existence of these back doors hasn't been proven, but all the same, the US embassy in Kabul and a military base in Fort Leonard Wood, Missouri, have now removed all their Hikvision cameras.

Recently the company attracted attention because it advertised a camera model on its website that could not only determine the gender of those photographed, but also "with a hit rate of at least 90 percent" automatically recognize whether the subjects are Uighur.[128] The *New York Times* had already revealed in April 2019 that China was the first country in the world to use artificial intelligence for racial profiling.[129] Concurrently, companies such as Yitu, Megvii, Sensitive, and CloudWalk have been working to improve their surveillance camera Uighur-recognition algorithms. According to the *New York Times*, such software is already being used in cities such as Hangzhou and Wenzhou, and in the city of Sanmenxia, 500,000 passers-by were screened over the course of a single month to determine whether they were Uighurs. "If originally one Uighur lives in a neighborhood, and within 20 days six Uighurs appear," CloudWalk said on its website, "it immediately sends alarms" to law enforcement. In October 2019, the companies named in the *New York Times* article were then placed on the United States "Entity List" for the human rights violations in Xinjiang. The blacklist severely restricts American trade with the companies concerned. For ambitious start-ups like SenseTime and Megvii, this is a setback in their efforts to expand worldwide.

The telecommunications company Huawei landed on the entity list a few months earlier. Huawei found itself caught up in the back door debate: Triggered by warnings from the USA, governments all over the Western world started discussing whether Huawei really should be used as a supplier for the coming 5G networks—the infrastructure of the future—or whether that was inviting the risk of Chinese espionage or even sabotage. In a series of interviews, the Huawei boss Ren Zhengfei assured everyone of his company's independence from the Chinese government. Huawei will "never cause damage to a nation or an individual," he said. What he didn't mention was that China's intelligence-service law from 2017 obliges all Chinese "organizations and citizens" to "support, aid, and cooperate with the work of the national secret service."

This document was merely a move by the CCP to

give a legal basis to something that had always happened, in a system that has never permitted companies to be independent in our sense of the word. "In China, state that Huawei strongly supports the Communist Party of China," says one internal notice from Ren to his executives, written in 2014 and quoted in the *Financial Times*. "Outside China, stress that Huawei always follows key international trends."

In the summer of 2019, a few months after Huawei was blacklisted in the United States, a memo from Ren was leaked to the Reuters news agency, in which he commits the workforce to a restructuring of global business and a persistent struggle for "life or death."[130] "After we survive the most critical moment in history," the Huawei boss wrote, "a new army would be born. To do what? Dominate the world."

Many other Chinese high-tech companies are penetrating overseas markets. In the province of Anhui, you can visit iFlytek, China's number one in voice recognition. "Today, we lead China," says a banner in the lobby, "tomorrow, the world." Another banner reads: "Let the machines listen and speak, let them understand and analyze. Let us build a beautiful new world with artificial intelligence." Like Huawei, iFlytek follows both the laws of commerce and those of the police state. It manufactures intelligent speakers, similar to Amazon's Alexa system. "Put it in your apartment," says the guide in the showroom, "and it will hear everything, no matter where you are." It provides voice-control technology for cars made by VW and Mercedes. Our guide points to large photos of Xi Jinping, who made a personal visit to the firm. "We have the support of the government," says Pan Shuai, the head of overseas sales. When the US president Donald Trump visited Beijing, a video clip produced by the firm went viral: "I love China," says a real-looking Trump in fluent Chinese: "iFlytek is really super!"

China has 800 million internet users and 1.4 billion mobile phones. iFlytek works with all three of the major mobile operators, running their call centers—and according to the company's own information, it has made its voice recognition software available to more than 370,000 apps free of charge. "Every day we have 4.5 billion users accessing our services," says Pan Shuai. "Our huge population gives us enormous advantages in data gathering." Of course, they don't break any laws, he adds, but: "China is the Wild West right now. A test laboratory."

And what about their cooperation with the policing ministry? Pan Shuai hesitates. "I don't know anything about that," he says. But it's on the company's website—

albeit only the Chinese-language version: iFlytek runs a laboratory for artificial intelligence and voice recognition jointly with the Ministry of Public Security. The company has developed keyword-spotting technology for public security and to aid national defense, the website says. They are also helping the authorities to compile a national speech-pattern and voice database.

iFlytek boss Liu Qingfeng is also a delegate to the National People's Congress in Beijing, and in that capacity he used a 2014 speech to call on the authorities "to employ big data in the fight against terrorism as soon as possible, in the interests of national security, and to begin compiling a nationwide speech-pattern database at once." The company's website says that they've already helped the police to "solve crimes" in the provinces of Gansu, Tibet, and Xinjiang. Its voice-recognition systems have also proven successful in the Tibetan and Uighur languages. After a while, the sales manager Pan Shuai recalls the firm's cooperation with the police. "Oh, right," he says, "that." Yes, it's true, the company is also making its technology available to the police. "We help with telephone scams, for instance." The technology is already in widespread use in the telephone networks of Anhui, the company's home province, according to a 2017 report on the Shanghai news website *The Paper*. It scans all calls in real time. "If our system recognizes a criminal, it alerts the police automatically," Pan Shuai says proudly.

iFlytek also landed on the American entity list in 2019 because of its role in high-tech surveillance of the Uighurs and the Xinjiang region. Like the face recognition company SenseTime, iFlytek also had a research partnership with the Massachusetts Institute of Technology; additionally iFlytek funded big data research at Rutgers and Princeton. Meanwhile, MIT has pulled out of the partnership.

There is a province in China that is a few steps ahead of the rest of the country in using all the surveillance technologies. It is a region that now has a police station every few hundred meters. In some districts you have to install a state-monitored GPS transmitter in your car, if you own one. You can only buy gas once your face has been scanned at the gas station and the system has declared you harmless. In every city, town, and village, cameras follow your every move, thanks to infrared technologies. If you've been identified as a potential troublemaker, then in some places the cameras will send an alert as soon as you stray more than 300 meters outside the "safe zone" that has been designated for you. If you own a mobile phone, you must install the Jingwang ("clean net") app on it. This app has access to the content of your phone and, according to the government, is supposed to "prevent people from accessing terrorist information." It detects all "damaging information" and "illegal religious activity" in the form of text messages, e-books, websites, images, and videos, and automatically reports them to the authorities.

At the countless police checkpoints you have to pass through several times a day in this province, officers scan your face with their smartphones, then check your phone to see if you really have downloaded Jingwang. In addition, some of the checkpoints are equipped with special technology that extracts information such as MAC addresses and IMEI numbers from cell phones and relevant devices without their owners ever noticing. If you buy a kitchen knife, a QR-code assigned to you will be stamped on the blade at the point of sale.[131] The authorities know how often you go to prayers, whether you have friends or relatives abroad, and whether you know anyone who has been to prison. All this is stored on your file, along with your fingerprints, your blood type, scans of your iris, and samples of your DNA, which the government takes at free health check-ups without informing you of what will happen to them. (The construction of the police force's DNA database relies on technology provided by the American company Thermo Fisher, as the *New York Times* recently revealed.) The sum of all this information determines whether you are permitted to stay in hotels, rent a flat, or get a job. Or whether you end up in one of the many re-education camps set up all over the province.

All this will apply to you if you're a Muslim living in Xinjiang Province, which in all likelihood means you're a Uighur. In 2016, the party leadership ordered a "People's War Against Terrorism" after a series of violent clashes between Uighurs and Han Chinese in Xinjiang and terrorist attacks in Beijing and Kunming, according to a massive trove of a leaked internal Chinese government documents related to the oppression of Uighurs (the "China Cables") published in late 2019 by the *New York Times* in collaboration with the International Consortium of Investigative Journalists. Xi Jinping personally ordered in a secret speech, published by the *New York Times*, that the "organs of the dictatorship" should show "absolutely no mercy."[132] Chen Quanguo, the new CP chief of Xinjiang, appealed to Xi Jinping and immediately after taking office in Xinjiang in August 2016, instructed security officials to "round up everyone who should be rounded up." Under Chen, the authorities started to destroy mosques and cemeteries. Anyone who calls their child Mohammed or

Fatima must expect to be reported to the police, as well as those who fast during Ramadan or who invite too many people to your wedding. If someone is too pious from the party's point of view (it is enough if he has grown a beard), if he emails or calls relatives abroad, if he has banned apps (such as WhatsApp) installed on his cell phone, there is a good chance that he will end up in a re-education camp. In fact the so-called Karakax-list, another internal Chinese government document leaked to Western media in the beginning of 2020, lists the following "crimes" as reasons for detention and re-education: "Has relatives abroad," "Has communicated with someone abroad," "Has a minor religious infection." Others were interned because they "had applied for a passport but did not actually leave the country," had a "complicated network of relationships," or merely because their "thinking is hard to grasp."

One of the most powerful tools of mass surveillance in Xinjiang is the "Integrated Platform for Joint Operations"—IJOP—an artificial intelligence system that collects data about all of Xinjiang's citizens and then, plugging that data into its algorithms, alerts the authorities about potential suspects. The platform comes with an app that allows police officers on the street to directly compare the facial and ID information of passers-by with the central database. The system's hunger for data is insatiable. Xinjiang officials should collect IJOP data "from everyone in every household," according to documents Human Rights Watch made public in 2019.[133] (The IJOP software had been leaked to HRW, which in turn had it broken down and analyzed by Berlin programmers using reverse engineering techniques.) The IJOP system is produced and operated by a subsidiary of the arms supplier China Electronics Technology Group Corporation (CETC). In 2016, CETC announced at a press conference that the company was developing big data systems on behalf of the government that would identify potential terrorists in advance.[134] "It's very crucial to examine the cause after an act of terror," said chief engineer Wu Manning. "But what is more important is to predict the upcoming activities." Two years later, the company signed a "strategic cooperation" agreement with the German Siemens group, which Siemens says opens "a new chapter in the digitization of China's electronics Information industry."[135]

According to the IJOP documents, anyone who deviates from what the party and its algorithm consider "normal" is considered a possible threat. The system reports as suspicious, for example, those citizens of Xinjiang who "keep a distance from their neighbors" or "who do not like to use the front door of their home." The same is true for people who move away from their registered place of residence without a police permit, who have "unusually high" power consumption, and who use a phone or a car that isn't their own. People whose phone goes offline longer, or whose GPS transmitter installed in the car suddenly no longer reports to the authorities, also become suspects. Additionally, those who have relations to suspects already under observation or "to people who are abroad" are classified as "problematic."

It is the sum of all this information that decides whether a person has access to hotels, whether he can rent an apartment or get a job, whether he is ordered to stop visiting public places, whether he is placed under house arrest, and whether he ends up in one of the province's many re-education camps.

"Document No. 14" of the "China Cables" reveals how the IJOP system identified 24,412 "suspicious" people in a single week in June 2017 in southern Xinjiang, of whom 15,683 were then sent to a camp for "training and education."[136]

Starting in spring 2017, China took just one year to create a gulag, into which (according to research by the German sinologist Adrian Zenz) a million or more people have already vanished—without having committed any crime. "Today, it's impossible to speak to a Uighur who doesn't have at least one close relative or friend in a camp," Adrian Zenz told me. "Any contact with people in other countries is almost a guarantee you'll be sent there." The government in Beijing denied the existence of the camps for a long time, but when satellite images proved otherwise, they adopted a new strategy: the camps, they explained, were in fact "schools" and "vocational training centers." Hand-picked groups of diplomats and journalists were taken on carefully stage-managed visits where they witnessed productions of colorfully costumed and happy Uighur "students" singing and dancing, and then in interviews expressing their deep gratitude to the Party for having saved them from the dangerous path of extremism. Former inmates from the camps who later fled abroad meanwhile continued to report about the draconian details of the re-education programs in which, day after day, they had to pledge allegiance to the Communist Party and Xi Jinping and swear off their own religion and culture. Some told of torture, rape, and forced labor. A police officer from Kuchar County actually told Radio Free Asia at the end of 2019 that 150 people had died in one of the camps outside Kuchar city over a six-month period.[137]

The publication of the China Cables in November

2019 finally proved the existence of the largest internment of an ethnic-religious minority since the Nazi
era.[138] The secret documents from the Xinjiang authorities described in detail the "schools" as sealed-off,
strictly guarded camps, fenced in with barbed wire
and watchtowers, and geared toward political indoctrination. Doors had to be "double locked" a document
instructed, and video surveillance was to be complete
and free from blind spots to "prevent escapes." The
documents speak of a punishment-and-reward system
and reveal that all inmates must be held in the camps
for at least one year. Prisoners are rated according to
a points system: plus points are given for "ideological
transformation," too many minus points automatically prolong detention. The task of re-education,
according to another government document from
Xinyuan County, is to "wash the brains and cleanse the
hearts, promote the right and eliminate evil."[139]

Not only has the Party turned Xinjiang into an enormous camp—it is also making the province into a testing ground for its AI gadgets. At the start of 2019 the
Dutch security expert Victor Gevers discovered a data-
set that the Shenzhen facial-recognition firm SenseNets
had parked, unprotected, online. It contained surveillance data on more than 2.5 million Xinjiang residents,
with their names and ID numbers. In a period of just

24 hours, these people—mostly Uighurs—had been logged at 6.7 million position points (mosque, internet café, hotel), evidently using data gathered from AI cameras.

Today Xinjiang is the "high-tech version of the Cultural Revolution," says Adrian Zenz. "The trend is toward automation, and predictive policing. Just as China is overtaking the USA, Xinjiang is already a long way ahead of the rest of China." Arrests are increasingly being carried out to fulfill quotas, claims Zenz. And in Xinjiang, the decision to arrest someone is increasingly made by technical systems; police officers don't check each individual case. Zenz sees a clear strategy at work: "the aim is to make citizens rush to obey, to internalize the controls, to self-censor."[140]

China's high-tech giants are getting in on the action. Huawei, for example, which for some time now has made half its revenue abroad, and supplies, among others, Vodaphone and Deutsche Telekom. In May 2018, the company started working with the Xinjiang police. It has since been involved in an innovation laboratory for the "intelligent security industry" being set up by the authorities in Urumqi, Xinjiang's capital. The stated aim of the regional government is to create an industrial park for high-tech surveillance technology—technology that can not only be put to use in the local area, but also exported to the countries involved in China's "New Silk Road" (or "Belt and Road Initiative"). After all, as the Urumqi government website explains, there is also a growing market in other countries for "anti-terror and stability maintenance products."[141] The site welcomes a "new soldier" to the battle for "a peaceful Xinjiang": Huawei will be a "close partner" in "technological" and "digitalized" police work. According to the report, the Huawei executive present at the opening ceremony for the innovation laboratory, Tao Jingwen—who at that point also happened to be President of the Western European Region for Huawei—explained that the company wanted to "bring the digital world into every organization, into every family and to every person." Without doubt, "public security is the most important element in this strategy." Together with the province's police force, they would "ring in a new era of smart police work." This is the same Huawei that abroad insists on its independence from the Communist Party, its representatives firmly claiming that the company does not do any direct business with the Xinjiang security authorities: "We remain in the commercial space," said Huawei's global cybersecurity manager, John Suffolk. This not only contradicts the cooperation in Urumqi described above; according to a new study by the Australian

Strategic Policy Institute (ASPI) of further Huawei activities in Xinjiang, the company also worked with the Karamay police on cloud computing projects and cooperated with the Aksu Prefecture Public Security Bureau modular data centers.[142]

In the preschools of Urumqi, all children under the age of six receive a free carton of milk, courtesy of the Party, every day. On the back of the carton are printed the verses of the song "Oh, Party, My Beloved Mother." At the same time, thousands of children are being placed in "welfare centers" because the state has put their parents in a camp: they are the re-education orphans.[143]

The first green shoots of a debate around risks and ethical implications have begun to appear. One chapter in a book published by Tencent's research institute, *Artificial Intelligence: A National Strategic Initiative for AI*,[144] discusses the question and calls for "strict rules." The State Council's plan has promised to introduce these rules in 2025, but as always in China, it is safe to assume that they won't stand in the way of the Party's need for control.

Public debate on state surveillance is taboo. Xu Bing's video surveillance montage *Dragonfly Eyes* makes a strong statement—but the artist ducks further questions, instead speaking in philosophical terms of a

Buddhist-inspired dissolution of identities ("Am I really me? And isn't everyone connected to everyone else?") or of his ideas about a "post-surveillance society." Orwell's dystopian vision? "Uninteresting. Everyone is familiar with that, anyway," says Xu Bing—who, incidentally, went to a great deal of effort to track down all the people who were recognizable in the images he used and asked permission to use them.

Industry executives like to argue that Chinese culture makes the country's citizens indifferent to data protection worries. "The Chinese are more open and less sensitive about data protection," claimed Baidu boss Robin Li at a forum in Beijing. "If people are prepared to exchange their privacy for efficiency—and people here often are—then we can make even more use of the data."[145] But is it really the case that the concept of "privacy" is alien to Chinese culture? Or is the disappearance of the term from public debate the result of Chinese politics? In any event, Robin Li's words triggered a surprisingly heated debate online. One of the most-shared comments called the Baidu boss "shameless and pathetic," and many others simply wrote: "I am *not* prepared!" In Hong Kong and Taiwan—which are also home to Chinese people—the topic of data protection and privacy regularly sparks flare-ups no less passionate than they are in Europe and the USA—and recently, tentative discussions have been on the rise in mainland China, too.

China's new cybersecurity law forbids private companies from misusing the data entrusted to them, though the same law places no such limitations on government authorities. Criticism of the state's frenetic data-gathering remains taboo, and online commenters almost always restrict themselves to criticism of companies and their excesses.

The cybersecurity firm Qihoo 360, for example, streamed the video footage from surveillance cameras it had sold to individuals and companies straight onto the internet, including live-streams from nurseries and school classrooms. The company said in its defense that it wanted to allow parents to check up on their children. After a tremendous online outcry, Qihoo 360 took the website down (it was one of the sites from which Xu Bing had taken footage for his film). And a survey conducted at the start of 2018 by the state broadcaster CCTV and the National Statistics Office—which questioned people between the ages of 16 and 60 in 100,000 households, including those in smaller provincial towns—found that more than 76 percent of those surveyed felt artificial intelligence was a threat to their privacy.[146] And at the end of 2019, law professor Guo Bing made headlines when he filed a lawsuit against a safari park in Hangzhou for "violating consumer rights": the park had started scanning all visitors with facial recognition cameras.

Nevertheless, the authorities do not look kindly on attempts to draw attention to the issue. When an artist in Wuhan went online and purchased the data of 346,000 citizens for the equivalent of a few hundred dollars—name, sex, age, telephone number, address, travel plans, online shopping—and then put them in an exhibition, calling it "The Secrets of 346,000 Wuhaners," the police closed the exhibition down and launched an investigation against him for data theft. And when someone dares to write a few words online about "mind control through artificial intelligence" as the essayist YouShanDaBu did on WeChat, the censors will delete it at once. Criticism of state surveillance remains prohibited. "The investigation of the entire population has become a reality," the essay said. "No one has a single corner left in which to hide. This is a beautiful era. Our only option now is to all think the same thought. Please give up all other thinking."

No need to worry, insists the Party paper *People's Daily*: "While some feel threatened by a technology that puts almost everyone under the spotlight, many more feel safer when such technology is in good hands."[147]

No need to worry, Xie Yinan from Megvii told me: "The data is all stored with the police. It's quite safe there." When he saw my skeptical expression, he added: "You want to talk about privacy? Do you have a smartphone? Well, then you've lost already. Your whole life is in there."

So, privacy isn't under threat, it's dead and buried—no point making a fuss about it.

The New Man

How Big Data and a Social Credit System Are Meant to Turn People into Good Subjects

"The trustworthy will be allowed to roam everywhere under heaven, but the discredited will find it hard to take a single step."

From the State Council's plan for setting up a system of social trustworthiness

China is once again focusing on the New Man. The good man. The honest man. "We want to civilize people," I was told by an official in the small city of Rongcheng, in eastern China. "Our aim is to normalize their behavior. When everyone behaves according to the norms, society is automatically stable and harmonious." The official beamed: "Then my work is a lot easier." The honest man is a trustworthy man. And trust might just be the scarcest resource in China.

China is experiencing a crisis of trust. No one trusts anyone. "Our society is still immature, and our markets are chaotic," says the Beijing professor Zhang Zheng. He points to the rural exodus and rapid urbanization of recent decades, which have been accompanied by huge social and economic transformations. "An honest man is a stupid man," said Wang Junxiu, describing China's social climate. He co-authored a report by the Chinese Academy of Social Sciences in Beijing, published in 2013, entitled "The Mental State of Chinese Society." In a survey by the Shanghai branch of CASS, 90 percent of those questioned said that anyone who was honest and trustworthy in today's China was automatically at a disadvantage.

Some people see the roots of the current climate of distrust in the political campaigns of the past, especially in the Cultural Revolution, when children betrayed their parents and wives betrayed their husbands in the name of their messiah, Mao Zedong. But Wang Junxiu also points a finger of blame at the present: "Fraudsters aren't being brought to justice, and the state itself is damaging the common good." In 2013, it was still possible to say this openly. That year, an essay written by the well-known Beijing sociologist Sun Liping made the rounds on the internet. It argued that "a country where even teachers and monks are corrupt is rotten to the core."

But 2013 was also the year in which Xi Jinping began his term as head of the Communist Party and the state. Xi recognized the challenge: a complete breakdown of trust was threatening to undermine the power of the Party, and at the same time it posed an obstacle to the long-term expansion of the Chinese market and continued economic growth. From toxic foods and environmental degradation to shoddy building work: for many entrepreneurs involved in the unbridled wild-west capitalism that dominates China, the pursuit of profit still justifies the most reprehensible practices. The CASS report came to the conclusion that society needed some kind of referee to ensure fairness. It recommended enhancing and reforming the rule of law; but the Party had other ideas. Instead it was going to do something the world had never seen before, and set up a "system of social trustworthiness." The Social Credit System.

I have come to meet Professor Zhang Zheng at Peking University. He is the dean of the Faculty of Economics, but more significantly, he's an important advisor for the new system. It's quite simple, says the professor. "There are two kinds of people: good and bad. Now imagine a world in which the good ones are rewarded and the bad ones are punished." A world in which those who respect their parents, never jaywalk, and pay all their bills on time are rewarded for good behavior. A world where these people are allowed to buy "soft sleeper" train tickets or given easy access to bank loans—and others aren't. Like the guy next door who cheated on a university admissions test, who downloads films illegally, or whose wife has just had one more baby than the state allows. It's a world in which an all-seeing, all-knowing digital machine knows more about you than you do. This machine can help you improve yourself by telling you, in real time, exactly what you can do to become a more honest and trustworthy person. Doesn't that sound like a fairer—a more harmonious—world?

Honesty. In Shanghai, they've got an app for that: it's called "Honest Shanghai." You just download it and register yourself. The app scans your face, recognizes you—and then brings up your life. As I learned when I visited the Shanghai Municipal Commission of Economy and Informatization, the app can currently access 5,198 separate pieces of information per citizen from a total of 97 public authorities. The Commission is where this information converges, and it is the Commission that is behind the app. Have you paid your electricity bill? Donated blood? Are you behind with your taxes? Did you travel on the metro without buying a ticket? The app stores your behavior and calculates whether the sum of all input is "good," "bad," or "neutral." Good Shanghai residents are currently allowed, for instance, to borrow books from the public library without paying the mandatory 100-yuan (about $14) deposit.

While the app is a gimmick to which people can sign up voluntarily, participation in the system behind it certainly isn't. Officially, it is called the "Social Credit System" (the Chinese title also translates as "system for social trustworthiness"). It was set to become a reality for every single person in China by 2020, even though effective nationwide implementation seems to be gradual and could take longer.[148] It already collects data on all residents of Shanghai. Shao Zhiqing of the Municipal Commission of Economy and Informatization takes pains to point out that his office doesn't evaluate people. In Shanghai, he says, this job is done by third-party service providers; the authorities merely forward the data to them. Their algorithms analyze and rate behavior as good or bad. Without doubt though, says Shao, the Social Credit System will change the face of China. "First of all, it will allow us to answer the question: are you a trustworthy person? It's all about bringing order to the market. We want everyone to stay within the law and abide by the contracts he has signed. The market economy is based on trust. And ultimately, what's at stake here is the ability to maintain order in society." In this system, all of us—individuals, companies, organizations—are nothing more than walking sets of data. And it is up to the government to harvest and evaluate all this endlessly flowing data. It enables the government to use incentives and punishments in order to control and steer our behavior as individual citizens but also as economic actors and as a society.

In Beijing, Professor Zhang Zheng is in good spirits—excited, even. Why don't we just get to know people better? We shouldn't be satisfied with just knowing about their past and their present, he says. "Looking at their future is more important." In Professor Zhang's words, the system aims to identify the bad people, companies, and civil servants. It refers to them as "the discredited," or "trust-breakers."

The goal is to find them out, at all times and in all places. With the help of big data, every citizen will get an evaluation stamp that will become their new identity, and which will ultimately determine how they live their life and what access they have to social resources.

Zhang Zheng isn't just a professor; he is also the faculty's Party Secretary—but that doesn't mean he's a fossilized ideologue. Zhang Zheng travels a lot: he's been to Japan and the United States. He's seen the world; he is inquisitive, clever, and at times also critical. We meet in a seminar room at the university. It's New Year's Day in the Chinese lunar calendar, the country's most important holiday. "Just listen," the professor says with a smile on his lips. "It's so quiet." It's perfectly possible that at that moment, we are the only two people having a meeting on the large campus. You can see that Zhang Zheng is passionate about his subject. "So, you're from Germany?" In Germany, he tells me, they also have an information system that allows banks and companies to check an applicant's creditworthiness. Does he mean Schufa, the German equivalent to the UK's Experian? Yes, Schufa and Experian, he says. Like them, but bigger. Much bigger. All-encompassing, in fact. "It goes without saying," the professor says, "that how you handle your finances is important. Whether you pay your debts on time." He looks me in the eye. "But let's also consider how you treat your parents and your partner, let's look at all your social behavior, whether you have a moral code. Doesn't that also provide crucial information about your trustworthiness?"

Then the professor mentions Rongcheng. "Go and take a look at what they're doing there. They're pioneers, no one else in China has got as far as they have."

With a population of 670,000, Rongcheng is a small city on China's east coast. It boasts a swan reserve, a nuclear power plant, and an Office of Honesty. "The people at the *chengxinban*, the Office of Honesty, are doing fabulous work," says the professor.

China's future is already being rehearsed here. Rongcheng is one of more than three dozen pilot projects in China. Each is taking its own path, but they are all trying to create honest men. "First, we need to let what we're doing here sink in with people," says Huang Chunhui, the head of the Office in Rongcheng. The Office of Honesty goes by another name these days, he explains, because "we realized the name was a little too vague." They are now the "Office of Creditworthiness." Huang takes a sheet of paper and draws an egg, slicing off the top and bottom sections with a stroke of his pen. This is society, he says. At the top, you've got model citizens. "And at the bottom, you'll find the people we need to educate."

Then he explains the system. It will cover every company and individual in China. Everyone will be continuously assessed at all times. In Rongcheng, each participant starts with 1,000 points, and their score can either improve or worsen. You can be a triple-A citizen (a "Role Model of Honesty," with more than 1,050 points), or a double-A ("Outstanding Honesty,"

1,030–1,049 points). But you can also slip down to a C, with fewer than 849 points ("Warning Level"), or even a D ("Dishonest") with under 599 points. And if that happens, your name is added to a blacklist, the general public is informed, and you become an "object of significant surveillance." This is how the Rongcheng municipality's handbook "Administrative Measures for the Trustworthiness of Natural Persons" describes it.

The "system of social responsibility" combines moral education and surveillance. With the help of big data, in the future both will happen in real time. According to a paper written in 2014, it is designed to eradicate "lies and deception"; it will "raise the honesty and quality of the nation" and promote a "harmonious society." And not least, it should also—in theory— make it possible for the general public to monitor the lower levels of government.

In Rongcheng, Huang Chunhui maintains that the system works, citing traffic lights as an example. In the past, he says, many drivers just didn't care: they would shrug and drive right through a red light, and then shrug again when they paid their fine. "No one dares to do that anymore, because they'd have points taken off their score." The first couple of residents whom I later approach on the street have never heard of "social trustworthiness." "What's that?" they ask me. A third seems unaware of the fact that the system is already recording and evaluating his entire life. "But actually, now you mention it, traffic rules are being more strictly monitored. That's why I drive so slowly. I think it's a good thing. A trustworthy society is a good thing, isn't it?"

The "First Morning Light" neighborhood is a stone's throw from the Office for Creditworthiness. There are neat lawns and new apartment buildings designed to house 5,000 families. The area has a total of 12,000 inhabitants. The streets are lined with VWs, Toyotas, and a few BMWs. This is how China's new middle classes live. "In the past, people knew no limits," Party Secretary Dong Jiangang explains. "Now, we're seeing the return of morality." As Party Secretary, he probably has to say this: after all, he is responsible for it. "We're building an honest neighborhood," he continues, pointing to a large board on the street outside his office, which everyone has to walk past. "We list the trust-breakers right here."

Ms. Wang let her dog do his business on the lawn and failed to remove it: a five-point penalty. Mr. Sun poured water onto the ground outside his building's door in winter, creating a patch of ice. He was also docked five points. "I'm sure they're ashamed of themselves," says Party Secretary Dong. "But that's just the way it is. Look at this." Now he's smiling. Mr. Zhou helped an elderly couple move into a new house; he earned five points. Mr. Li gave a calligraphy class: five points. The Yu family offered the community its basement for a singalong of Communist songs from the Revolution: five points. People in this neighborhood can also earn points by shoveling snow, taking elderly people to doctor's appointments, or helping children with their homework.

Dong Jiangang then points out his star residents, those with more than 1,000 points. They have received "Honest Family Role Model" awards for their good behavior. Qin Zhiye is one of these model citizens. The retired 64-year-old is a Party member. Visitors to his new apartment are invited to sit on the sofa, which he has neatly covered with a pink rose-patterned quilt. Qin has lived in the "First Morning Light" community for seven years: "I've developed feelings for every blade of grass and every tree here." And of course, he says, the system has improved community life. "If you lose a lot of points, people start whispering about you: 'Look at him, he's a B. Or a C.' That's a shameful thing. Sometimes they just have to warn somebody: 'Hey, we're downgrading you.' Then people get scared."

The whole neighborhood is divided up into "cells," with 400 families to each unit. "These families keep an eye on each other," says Dong Jiangang. "We've got people who inspect their block. They sometimes question other residents, and they take pictures and record videos of bad behavior."

The Party began to roll out this technology-assisted "grid management system" of monitoring in 2001, in cities like Shanghai. There is a long tradition of these systems in China. In the fourth century BC the statesman Shang Yang, who laid the foundations for China's unification and thus paved the way for the rule of its first emperor, divided the population into groups of five to ten families. People were required to monitor and denounce each other. If one person made a mistake, everyone in the group was punished. One document states: "If the punishments are severe and collective, then no one will even dare attempt [to commit a crime]. And if no one even tries, there is no need for punishment." Already at that time, autocrats were dreaming of the new man who has internalized control, the subject who monitors himself.

The Communist Party's moral and social credit system also has historical precursors. The 16th and 17th centuries saw a great demand for "ledgers of merit and demerit" among the population of China. They were influenced by Confucian and Buddhist moral ideas and ways of life; anyone who was in doubt about how karma would reward their deeds, both good and bad, could simply look it up in these ledgers. It was a time of political and economic upheaval, social and moral confusion. Commercialization and the growth of the market brought the old hierarchies into question, and in some cases turned them on their heads. In a society presided over by Confucian scholars, the merchant class had always been the scum of the earth, but all of a sudden the merchants started climbing the social scale, gaining prestige and influence.

The ledgers gave confused and helpless citizens something to cling to in an age of state corruption, changing values, and rapid mobility. In Yuan Huang's ledger, for example, anyone who saved another person's life or a woman's chastity was awarded 100 points. Ensuring the continuance of the family line or adopting an orphan was worth 50 points, and recommending a virtuous person for office was worth 10. This is not so different from the ledger of the First Morning Light neighborhood. It's just that today the CCP and big data have replaced karma, and reward and punishment will come your way in this lifetime—ideally here and now.

In the city of Rongcheng, the municipal handbook says that the system aims to provide "incentives" and that punishments are only meant to "help." Dong Jiangang, the Party Secretary for First Morning Light, is thinking about the bigger picture, far beyond his neighborhood. "We are ensuring that society is harmonious." Dong tells me about the parents who come to him before their daughter's wedding, eager to find out about their potential son-in-law's score—the system already serves as a son-in-law trustworthiness meter of sorts. On his rose-patterned sofa, old Qin lets out a laugh. "Of course you'd want to know what you're letting yourself in for!" says Dong Jiangang. "Admit it: you would do exactly the same thing."

Being a model citizen also makes financial sense. "If a person has a lot of points, like our fine Mr. Qin here, then they no longer need to provide guarantees when they want a bank loan," Dong says. "Isn't that great? But that's how our Party works: if you're good, it will be good to you." What if I'm bad? "Then eventually you won't be allowed to board planes or high-speed trains. And I won't hire you."

It's true. This is exactly what it says in the document entitled "Warning and Punishment Mechanisms for Trust-Breakers," published by the CCP's Central Committee and the Chinese State Council in October 2016. Companies that engage in fraudulent activities risk being excluded from public tenders. Citizens who have lost trust points aren't allowed to apply for government jobs. Their access to insurance and bank loans is either limited or removed altogether. Buying a car or building a house becomes difficult. People with low scores can no longer fly, buy tickets for high-speed trains, or travel in the comfortable cabins on sleeper trains. Their internet access is restricted to low-speed networks. Trust-breakers don't get to stay at luxury hotels or eat in exclusive restaurants. They can even be prevented from traveling outside China, and their children are denied access to the country's expensive schools.

Many details of the system are still unclear. Which of the three dozen-plus regional models will win through? Can they continue to exist in parallel, or will unified national standards be set out? Will the country end up with a kind of social-credit ecosystem, housing a huge range of incentive and punishment models, or one seamless machine? When and how will the credit system be integrated into the private-sector economy? Officially, the system was supposed to launch nationwide in 2020, but at the moment it looks like the rollout will be gradual and over the next few years. At the national level alone, no fewer than 46 Party organizations and government bodies are working out the details. The government is in the process of introducing a single, country-wide credit number system: in future, each citizen and each company will be allocated a number, under which all relevant information will be saved.

A few pieces of the puzzle are already in place: for decades, China has operated incentive systems that reward morally upright, conformist behavior, turning grafters into "heroes of work" or families into "role models of hygiene." In 2013, a law was enacted requiring all Chinese people to care for their aging parents under threat of punishment. The same year, China's Supreme Court began keeping a list of defaulting debtors, which became the basis for the Social Credit System's nationwide blacklist. The National Public Information Center has now announced that in 2018, seventeen and a half million Chinese citizens were denied access to planes, while five and half million were not allowed to buy tickets for the high-speed trains. However, the number of punishments is increasing rapidly: according to the *Global Times* newspaper, in the month of July 2019 alone, 2.56 million trust-breakers were already prevented from buying a plane ticket. In two districts of Henan Province, the courts are working with local telecom companies: if you make a call to someone on the blacklist, then instead of the ringing tone, you'll hear a message informing you that the person you're trying to reach has been placed "on the list of trust-breakers."

Public exposure is built into the system. The internet portal Credit China, which the government has set up for the credit system, has posted a series of cartoons by way of illustration.[149] If the courts have put someone on their blacklist, the website says, "then his photo will appear on large screens all over the city, so that he will no longer be able to hide and passers-by everywhere will be able to see his photo and his personal information."

One of the cartoons shows a young man in a suit and bowtie brandishing a bunch of roses for the woman he adores. He is breaking into a sweat because he has evidently just been identified as a defaulting debtor. "Disgrace" says a box above his head, with an arrow pointing to him. And the pretty girl declines his advances: "You never pay back your debts," she says. "I've seen your photo on the street committee's big video screen. No one will want to go on a date with you now." The caption warns: "Don't ruin your whole life with your breaches of trust."

The music video app Douyin (TikTok, in English) has begun to involve the public in the hunt for trust-breakers—another example of the close cooperation between private internet companies and the state. Users in Guangxi Province were shown images of wanted people who had been placed on blacklists between their music videos. Anyone who knew where to find one of these people, the app said, should tip off the police. As a reward, they would receive a share of the value of that person's debts.

Credit China allows you to search for specific people whose names appear on blacklists. The virtuous are placed on "redlists." A second website provides similar information about companies that have been sanctioned: there were 3.59 million of them in 2018, among them, to take just one example, scandal-hit vaccine manufacturers. The social credit rating system will also affect foreign companies. A study by the EU Chamber of Commerce in Beijing warned all companies to prepare well, as the system already was a reality and, according to chamber head Jörg Wuttke, "could spell *life* or *death* for individual companies."[150] You can already see companies' scores online, calculated from past behavior in categories such as labor law, environmental protection, hygiene, or contract compliance.

The mechanisms for punishment, according to an official notification from March 2018, work on the principle that "breaches of trust in one place bring sanctions everywhere." So a flying ban might be imposed for air rage, but also for tax evasion, financial misconduct, or underpaying social security contributions. In future, month-long train-travel bans will be issued not only to people caught illegally re-selling tickets, but to anyone who lights a cigarette on a high-speed train.

The *dang'an*, the secret file that documents the lives of every single individual, is a relic from the days of Mao. It includes information on people's preferences and aversions, their career path, and their political reliability. In the past, this information was only kept in paper files, which in recent years had almost been forgotten. But now, all the individual pieces of data are being pooled into a new whole. The official guide to the system contains a section entitled "Acceleration of the Punishment Software," in which its aim is described as "automatic verification, automatic monitoring, and automatic punishment" of each breach of trust. No more loopholes, anywhere.

Although the government has been working flat out on the system since at least 2014, most Chinese have never heard of it. "Well," a friend in Beijing shrugs, "in this country, we're all naked anyway." When it comes to access by the Party and the state, hardly anyone expects real privacy in China, and most people are fatalistic about it. After all, you can't be more naked than naked. Unless, say, someone opens up your skull and peers into your thoughts.

The existence of the system is only gradually seeping into public consciousness as more blacklists of trust-breakers appear and more actual cases of reward and punishment are published in the Party media. The Beijing subway said at the end of 2019 that it wanted to install new facial recognition cameras and then integrate them into the system.[151] The cameras are supposed to identify passengers whose names appear on blacklists as a possible "security risk" for stricter controls. Anyone caught eating on the subway will get a point deduction. Also, in 2019 all blacklisted citizens were for the first time excluded from participating in the state exams for admission to the civil service. "The trustworthiness society is nearly here: are you ready?" asked the *People's Daily*. When I have told Chinese people a little about the system, a lot have reacted positively, saying that it would be great to finally have more trust within society. But when in spring 2018, a state middle school in Shandong Province announced that from now on it would not be accepting applications from children whose parents were on blacklists, there was a storm of online outrage at this visiting of the sins of the parents upon their innocent offspring.

The best-known example of a social-credit pilot project in China is not state-administered, but run by a private company. Sesame Credit is part of the Alipay app, the market leader in cashless payments (Alipay claims to have 1.2 billion customers together with its partners). All Alipay customers are able to activate Sesame Credit, and are then rated on a scale between 350 and 950 points. The algorithm that evaluates the data is secret. But the program's creators have clearly specified five areas on which your rating is based: in addition to your identity, your ability to pay back debts, and your financial history, behavioral preferences and personal networks also count.

Li Yingyun, the project's technical director, has said that, for example: "Someone who plays video games for 10 hours a day would be considered idle, and someone who frequently buys diapers would probably be recognized as a parent, who is more likely to have a sense of responsibility." According to the Shanghai website *The Paper*, people can lose points for changing addresses too often. "What's more," *The Paper* writes, "your friends' scores will affect your Sesame rating." The message is clear: stay away from friends with low scores.

Sesame Credit cooperates with Baihe, China's largest dating platform: singles looking for a partner can advertise their Sesame scores there. Point champions will also have easier access to loans, and they already enjoy an express visa service to countries such as Luxembourg and Singapore. Alipay is increasingly looking to expand into the West. Chinese travelers can already use Alipay in many places in the United States or Europe. The trade-off is that the app then transmits their transaction and GPS data back to China in real time.

Even so, the future of the commercial credit systems is uncertain. The Chinese People's Bank granted eight private companies permission to run pilot projects, but recently decided not to issue any with a permanent license, citing data protection reasons, as well as the fact that "conflicts of interest" could not be overlooked.

Many have argued that Alibaba is using the incentive system of its Sesame Credit subsidiary mainly to boost its own revenue rather than the social behavior of its users. Reports have also started to emerge about how the system can be gamed: in one article on Weibo, hackers boast of people paying them to tweak their data on Alipay so that it generates more points. A report by MERICS predicts: "As the Social Credit System unfolds, it is likely that additional methods of data forgery will arise."[152]

The system's creators emphasize the benefits of "social trustworthiness." They say it helps rein in the worst excesses of the fraud that allegedly costs China $90 billion a year. Electronic payment systems will also, they predict, give millions of underprivileged

Chinese citizens access to credit. In the past, groups such as farmers, laborers, and students were unable to prove their creditworthiness. Ultimately, of course, the project is designed to trigger consumption and deliver a massive economic stimulus.

Some people are not so sanguine. The writer Murong Xuecun finds the system "creepy": "They talk about trust. In reality, it's about control, down to the very last personal detail. What will happen if you write something stupid on the internet as a young man? Okay, they may not arrest you, but they might confiscate your passport or your driver's license. Or freeze your bank account."

Since Murong Xuecun's Weibo accounts were shut down and he was banned from publishing novels and screenplays, he has been scraping a living selling cosmetics, strawberries, and honeydew melons (among other things) on the internet. "Once the system is really up and running, China will surpass anything George Orwell dreamed up," he says. "As a politically unreliable oddball, I will probably get a very bad rating. It might stop me from traveling abroad or taking the train. Maybe my landlord will throw me out. Well, I guess then I'll have to sleep under a bridge." The author was more than an hour late to our interview. The secret police had read about his appointment with me on WeChat, and tried to persuade him to cancel until the very last minute.

Some take comfort in the idea that the new surveillance system may never work as it is meant to. After all, it comes with huge technical challenges: pooling all the data, ensuring its accuracy, evaluating it effectively . . . The Communist Party has a long history of failed attempts to "civilize" the Chinese population. Despite countless campaigns, people in Beijing continue to spit on the street, and some residents of Shanghai still walk to the grocery store in their pajamas in broad daylight.

But would it really be better if the system turns out to be rife with the inefficiency, manipulability, and corruption that currently characterize the Chinese state—and yet still had the power to reward and punish citizens? The system's creators have two central aims: trustworthiness and political control. And while it is entirely possible that the first will fail, the latter is likely to work perfectly.

In the end, the Party still decides who is trustworthy and who isn't, and why. A person in Rongcheng who engages online in "illegal religious activities" (in this part of China the phrase mainly applies to the harshly persecuted Falun Gong movement) loses a whopping 100 points. It's the maximum deduction. Petitioners who try to draw attention to injustices they have suffered at the hands of local authorities during important Party or government conferences are docked 50 points. Anyone who takes their petition all the way to Beijing is automatically reclassified as a Category D: dishonest trust-breaker.

In Rongcheng, you can also be punished for "negative online behavior": say, for posting a comment on Weibo that could have "a damaging influence on society." Who decides what is damaging or negative? "Don't worry," says Huang Chunhui, the director of the Office of Creditworthiness. "We only act on things that are uncontested and have been proven beyond the shadow of a doubt." Through court rulings? "Not just that," Huang says. "We also rely on the assessment of the security services." In the end, it's the police and state security who make the call.

The system doesn't just capture the "social trustworthiness" of individuals, but also of companies and organizations—in fact, of every organization operating in China, meaning that foreigners are also affected. Here, too, the political direction of travel is becoming ever clearer. Non-governmental organizations (NGOs) in China had been struggling for air after a law passed in early 2017 to bring them under control.[153] Then, in 2018, some were issued with a 40-page handbook, the main message of which was that they—and their leaders—must submit to the social credit system with immediate effect. In future, they would win and lose points for every move they made.

"Endangering China's reunification and national unity?" A fat 100-point deduction for the foundation, and 50 for its CEO. "Libel" or "publishing damaging information"? Also 100 points docked. Is any criticism of Xi Jinping and the Party now taboo? How about expressions of sympathy for democracy in Taiwan? And would, for example, the Heinrich Böll Foundation in Beijing face penalties if one of its representatives were to appear at a Tibet or Taiwan event in Berlin? Is the CCP hoping to use the system to export its values and political ideas? The handbook suggests as much.

Of course, your organization can also gain points: for instance, by promoting "international friendship" (5 to 10 points) or by forming "grass-roots cells of the Communist Party of China." A CCP cell within the Konrad Adenauer Foundation, which is affiliated with the right-of-center Christian Democratic Union party in Germany? It doesn't seem likely to happen any time soon.

Zhang Zheng, the economist and professor at Peking University, raves about the advantages of the system. It's an opportunity, he says, to deal with badly-run companies, doctors who take bribes, violent teachers, and corrupt civil servants. However, he is also one of the few who warn of the dangers, which include a disproportionate concentration of data and abuse of state power. This is precisely why he is against the creation of a central database, into which all other databases in the country would feed information. This all-knowing database, if ever created, would be the new brain of Chinese society.

The powerful National Commission for Development and Reform is overseeing the introduction of the system, and is currently weighing up the idea of a central database, Zhang Zheng tells me. "We think that's dangerous," he says. Who does he mean by "we"? "Academics," the professor says. But, he adds, there are "also other voices." These would be the voices of the security apparatus—and in a regime like China's, they may carry more weight than academics. "Our biggest problem is that we don't have anything to model our system on. We're breaking new ground," says the professor. "But this is precisely what makes the whole thing so exciting." He leans forward, almost overcome now by his enthusiasm. "There has never been such a system in the history of humankind. And it still doesn't exist anywhere else in the world. We're the first. It's exciting."

The first. It will serve as a warning to all democracies

where companies and public authorities are dreaming their own big data dreams. And, no doubt, as a temptation to others: according to the Berlin-based Mercator Institute for China Studies (MERICS), "China's IT-supported authoritarianism will become an attractive role model and technology provider for other authoritarian countries." The former MERICS chief Sebastian Heilmann calls it "digital Leninism."[154]

We are witnessing the return of totalitarianism in digital guise. The People's Republic of China has always been a dictatorship. But it was only for a few years under Mao that it was a totalitarian state, which tried to creep into every last corner of its subjects' brains, its eye watching over their bedrooms and their closest relationships. The new totalitarianism will be much more sophisticated than the versions that Mao and Stalin gave us, with undreamed-of possibilities for access and mind-control, now that we have all stored our minds in smartphones—now that we record every step we take and every thought we think digitally. Best of all, the new totalitarianism has the luxury—unimaginable in the past—of being able to dispense with terror as an everyday tool. It's enough if the violence remains at a subliminal level, as an ever-present threat. In this way the new regime insinuates itself, quietly and imperceptibly at first, making citizens into its accomplices.

"Wouldn't it be the best of all worlds if, in a few decades, we didn't have to talk anymore about the system and its rules?" Zhao Ruying asked me. She is the department head in charge of implementing the Social Credit System in Shanghai. "We may reach the point where no one would even dare to think of committing a breach of trust, a point where no one would even consider hurting the community." She beamed with delight at the thought. "When we reach this point, our work will be done."

Then the new man will have been born.

The Subject

How Dictatorship Warps Minds

"How easily we can become slaves, and still be
very contented with our fate."

Lu Xun, 1925

S ome people are losing the plot.

A day laborer, in the interior, somewhere near
Chongqing. One of tens of millions. With a son who is
a stranger to him, and an even more estranged mother.
He gets his hands on a gun, steals a motorbike, robs a
bank. Someone gets in his way. He reaches for the gun.

A woman in the heart of China, in Hubei Province.
Her lover is married; his angry wife beats her up. She
works in a sauna. A customer hits her with wads of
cash, humiliates her. She pulls a knife.

A young migrant worker in a southern city. Dongguan. With no money, no friends (he's betrayed them), no family (his mother despises him for having made no money), no love (the woman he loves offers herself to other men in a brothel, right in front of him). He climbs up to the top floor of his hostel (which is called "Oasis of Growth") and jumps out of the window.

A laborer in Shanxi, coal-mining country. The bosses once promised to share the mine's profits with the workers. Now they're pocketing the money themselves. The laborer wants justice and instead receives a public beating, in front of everyone. The mafia-like bosses spit on him. His colleagues, though they have all been betrayed just like he has been betrayed, laugh at him. He gets hold of a shotgun.

There is a great deal of bloodshed.

These are dark tales. But, says the storyteller, they are "like a match that shines a light on us," dispelling the shadows of oblivion and denial. The torch-bearer is Jia Zhangke, and his film is called *A Touch of Sin*. It's a feature film rather than a documentary, but the violent acts he describes are all based on real events in China. The young man really jumped, the woman really stabbed someone. All that blood really flowed.

It was the internet that washed all these stories up at Jia Zhangke's desk. The director fished them out of Weibo's mighty river just before the Party closed the floodgates of censorship in 2013, assembling them into a grand panorama. In traditional Chinese art, such landscapes—grand, bold, expressing something about the state of the nation—are known as paintings of "rivers and mountains, ten thousand miles across."

"It's something I always wanted to do," says Jia Zhangke. "Make a film that summarizes the current state of China." It is one of the best films to come out of China in recent years—and one of the gloomiest. *A Touch of Sin* shows a new emptiness behind the new prosperity. A country without justice, a society without morals, people who turn to violence because they think it's the only way they can retain a last shred of dignity. The film borrows from *wuxia*, the Chinese martial-arts genre, in terms of both its aesthetic and its storytelling. The classic *wuxia* novels with their robbers and rebels, the films with their fearless heroes, were fables told by desperados who reached for their swords when they were forced to the edges of society, when there was no other escape. But Jia's heroes are broken, with no hope of comfort or catharsis.

Jia Zhangke grew up in Shanxi Province, a soot-covered coal-mining area. He was the son of a teacher and a shop assistant, in a very ordinary neighborhood. First he studied art, and in 1993 he went to the Film Academy in Beijing, where he studied the theory of film and greedily devoured the entire history of the medium. "The Academy was a place of freedom, with no taboos," he says. "I arrived and the first thing I did was borrow a Japanese porno. It was a happy time, it freed me." Like other filmmakers of his generation, Jia Zhangke deliberately set himself apart from role models like Chen Kaige or Zhang Yimou. The young director adopted a new realism, in contrast to the opulent images and historical parables of the filmmakers of the older generation. "Films of today should tell stories of today," he says. After all, China is bursting with them. Xi Jinping allegedly once said that he liked Jia Zhangke's laconically realistic films. Mind you, that was before *A Touch of Sin* was released.

Jia Zhangke is sitting in his office in the north of Beijing, with Mao Zedong on the wall behind him: the French poster for his film *Platform*, released in 2000. On the original poster, the image of Mao is upside down, but Jia has turned him the right way up. He talks about the culture of violence with which many Chinese people have grown up, thanks to Mao. In a different China, his film wouldn't have had to be an outpouring of violent outburst (he could have made courtroom dramas, he says), but he wants to show the damage done by this period of history. A society that has lost empathy and humanity. Corruption and self-enrichment by the powerful. People without hope, humiliated, who resort to violence not because it promises a solution, but as a final act of self-determination. "It's the only way they can tell the story of their despair," says Jia Zhangke. "They're under great pressure, and all other channels, all other vents, are closed to them. That's dangerous."

In truth China isn't being overtaken by an epidemic of murders and attacks—the country is probably still one of the safest in the world. Yet there are regular, disconcerting outbreaks of violence: a wheelchair user who detonates a bomb in Beijing airport; an impoverished street trader who blows up a bus in Xiamen, killing 47 people; a dozen petitioners from the provinces who swallow poison together near Tiananmen Square, because collective suicide is the last available means of protest. Something is festering beneath the stability so often invoked by the Party.

"In the past, before Weibo existed, when people heard these stories they thought they were one-offs," said Jia, when I visited him in the summer of 2013. "But since we've had Weibo, we can see that these things are happening all the time, all over the country. They're part of our reality, there's no denying them any longer." Jia Zhangke's film was a window. Just as Weibo was a window, for a short while. Not long after my visit, the authorities informed Jia that his film, which the censors had already rubber-stamped, could now not be shown in the country's cinemas after all. A new wind had begun to blow. At the same time, Xi Jinping's people put the shackles on Weibo. Denial was back.

It is China's dramatic transformation, and the fate of ordinary people who are losing themselves and each other as a result of it, that haunts Jia Zhangke. Some critics see Jia as an activist for the "underclass," but the director dislikes this characterization: "It's funny how everyone in this country, no matter whether they're farmers, workers, or intellectuals, always talks disparagingly about the 'underclass,' and how none of them think they belong to it themselves. But in China we all live under an authoritarian regime. That means we're all the same. There are those to whom the power belongs—and then there's everyone else. We're all the underclass."

"Blissful ignorance"—the ideal state in which an autocracy strives to keep its subjects. They submit to a policy of infantilization, which treats adult citizens like small children, incapable of thinking or taking responsibility for themselves; always prepared to be chided and educated for spitting in the street, line-jumping,

talking loudly in public places, insolence and disobedi-
ence. For years the censor was successful in keeping
much of the Chinese population in this state of ignorance. Then the authorities made the mistake of allowing wild and free debate to flourish on Weibo.

This accidental four-year gift of freedom was all it
took for cracks to appear in the picture painted by the
propaganda machine. A few courageous intellectuals,
artists, and writers looked behind the glittering façades
and the new prosperity, to spy an entirely different
society. "When I look at contemporary China, I see a
nation that is thriving yet distorted," wrote the author
Yan Lianke. "I see corruption, absurdity, disorder, and
chaos. Every day, something occurs that lies outside
ordinary reason and logic. A system of morality and a
respect for humanity that was developed over several
millenniums is unravelling."[155]

Almost everyone I spoke to in the first two years after
my return to China in 2012—friends and interviewees,
typical representatives of the new urban middle classes
who had profited from the Chinese miracle—had one
wish: to get out of China. There was a surge in searches
containing the term "emigrate" on Baidu. "One hour
on Weibo is enough to make you depressed for a week,"
a Beijing friend told me. People thrilled at the flood

of information on Weibo. And soon, they despaired. A sense of hopelessness began to spread. So this was the future they had been promised?

It was at this time that Xi Jinping took office and revealed his "China Dream" of the rebirth of the glorious nation. But people all over the country were finding common ground for the first time through the new social media—and they were telling each other about their own personal dreams. "We used to dream the dreams of our state," a prominent academic told me. "But now people are saying: these dreams have nothing to do with me." He talked about his daughter, who was studying English. "My dream is for her to have a good future."

These were precarious years for the Communist Party and its project of limitless and eternal rule. Large sections of society were beginning to wake from the hypnotic trance in which Party propaganda had kept them for decades. There were three sentences I heard time and again:

Mei you anquangan.
"There's no feeling of security here."
Mei you xinren.
"There's no trust here."
Mei you daode.
"There's no morality any longer."

There are brave, passionate, sympathetic people in China—people with a sense of solidarity, who don't hesitate to help others. But this isn't as easy as it is in some other societies; such people are often viewed with suspicion by their fellow citizens. They may also find themselves in the cross-hairs of the state apparatus. For example, in the cold winter of 2017 there was a pitiless campaign to drive non-Beijingers out of Beijing, when the city authorities literally took the roof from over the heads of many migrant workers. A lot of Beijing citizens were shocked and outraged. Some offered help to those who had been made homeless overnight—only to feel the might of the security apparatus themselves as a result. Yang Changhe from the district of Tongzhou, who gave the homeless a room in a Beijing suburb to store their possessions, got a visit from the police, who then made sure that he lost his own apartment. The painter Hua Yong, who captured the forced evictions on film, had to flee Beijing and was hunted for days before being arrested in Tianjin.

To give another example: the first time the civil rights activist Hu Jia was arrested, it was because of his efforts on behalf of the sick in the AIDS villages of Henan Province. "Why is it so often the good people who are targeted in this country?" asked Hu Jia. Here is the logic of a corrupt regime, which views all kinds of idealism as suspect.

Every autocracy seeks to destroy solidarity and sympathy among its citizens. Today's China is seething with mistrust, and complaints about moral decline have reached a new pitch of intensity. There are several reasons for this. One is the trauma of the Cultural Revolution, which is still unhealed. Another is the vacuum where faith in communism used to be. When that went to its grave along with Mao, people were left with nothing they could believe in. There was only the Party's rallying cry: Get rich! Greed is the imperative in this new China.

"My generation has no values and no principles," said Charles Zhang, who was born in 1964 and is now the head of the internet company Sohu, which makes him one of the stars of the new China. "And so the laws of the jungle rule. You can be successful for a whole range of reasons here. Instead of sticking to your principles, you start just exploiting everything. You exploit to a huge degree."

When the market research organization Ipsos conducted a global survey in 2017 on what worried people most, the majority of countries placed unemployment, corruption, and inequality at the top of their list.[156]

China was the only country where people's top concern was "moral decline." Life under an autocracy corrupts society, poisoning individuals and their relationships with others. "Wherever it has ruled, it has begun to destroy the essence of man," Hannah Arendt writes of the totalitarian state.[157] China's society is sick, even if at first glance that isn't always apparent to outsiders.

The law of the jungle. The worship of money and power. A brutal pragmatism, distilled from those long years when basic survival was the only value that counted. One of the most remarkable phenomena during Donald Trump's election campaign and the first months of his presidential term was the growth of his fan club in China. Trump's anti-Chinese diatribes were far less important to these fans than his shameless campaign against all forms of political correctness. They applauded his attacks on the welfare state and other countries' refugee policies. In their minds, these things were the work of misguided and despicable Baizuo.

Baizuo are "white leftists." For a while the word was a popular insult among certain Chinese online communities. It's what they call Western liberals like Hillary Clinton or Angela Merkel (for whom the sarcastic term shengmu, "holy mother," was also invented because of her stance in the refugee crisis of 2015)—and, more broadly, anyone who speaks up for asylum-seekers, the underprivileged, or LGBT rights. Such people are branded hypocrites and weaklings. Everyone is responsible for his own life, the mantra goes, and for getting himself out of difficulties.

In a 2018 study, Norwegian researchers were surprised to discover that, among the Chinese people they had surveyed, the proportion who held right-wing libertarian ideas, accepted social inequality as the natural order, and opposed higher taxes on the wealthy was far larger than it was in the USA. One in five Chinese people thought the state should stop all redistribution of wealth. "Chinese are more right wing than Americans," *Science Nordic* announced on its website.[158]

"Many Chinese people don't hate the rich; they want to be rich themselves. Instead, they hate morality. So much has gone wrong here," Li Chengpeng told me in Chengdu. The former sports reporter and liberal bestselling author is among the celebrity social critics silenced in the 2013 campaign against independent bloggers and authors. Social Darwinism, which transposes the ideas of natural selection and survival of the fittest from the natural world onto society, has long since taken hold in China, says the historian Cheng Yinghong: "Social Darwinism and autocratic politics are natural bedfellows. It ignores the injustice inherent in the unequal distribution of rights and resources.

And it protects the violence, deceit, and exploitation on which these things are based."[159]

"We live in an age when dust blocks the sky. Politics is dirty, the economy is dirty, and even culture smells like it's rotten," the author Murong Xuecun writes.[160] "Our heart is supposed to be clear like the water in the autumn and the unending sky, but if we place it in the dust for a long time, then it can't help but get dirty and frangible." This is an essay written in the tradition of the great Lu Xun, and it's one of the most clear-sighted analyses of modern Chinese society around. On the day it was published, it was one of the most shared posts on Weibo, and of course it was deleted as quickly as it appeared. Murong Xuecun describes the devastation that an autocracy wreaks in the souls of its subjects. He identifies several states of mind, foremost among them numbness, blindness to reality, and slavish subservience. All are the result of generations of people for whom basic survival has been a consuming struggle.

Here, for example, is Murong Xuecun on the citizen who has entered a state of numbness: "If someone steals his food, he just goes hungry. If he is slapped in the face, he accepts it. If his home is destroyed, he watches. If his baby is terminated against his will, he simply cries. Any injustice is accepted as unavoidable; as expected, even. It would be abnormal if things were any different." The numbness also applies to other people's fates. "If people are beaten, he just stands and watches. If people complain, all he can do is mock them. When people announce that they are going to commit suicide, he says: 'Oh, they just want to be famous.' If no one speaks up for him, he accepts it. But if someone does, then he thanks the fates and says 'Hey, that's a stroke of luck!' If someone speaks up for him without success, then he pretends he always knew that this would happen: 'A waste of time!' And if someone speaks up for him and is carted off by the police for it, he giggles and says: 'Serves him right, the troublemaker.'"

When the numbness reaches an extreme level, the subject starts to hate everything that is good and fair. He thinks all idealists are hypocrites, and anyone who stands up for justice is a shameless opportunist, really just pursuing his own interests. A sense of moral inferiority triggers a defensive reaction in people everywhere. In the West, too, studies have shown that when one person in a group does the right thing, morally speaking, they attract hatred rather than admiration from their peers.[161] Their actions show the others their own flaws and remind them that a different approach is possible. This dynamic can be observed in all societies, but it is especially noxious in places where suspicion of other people's motives has become a matter of principle.

A hundred years ago, Lu Xun wrote about how readily subjects living under authoritarian regimes become slaves—out of fear, opportunism, or numbness. "But today, most people don't think they're slaves; they think they rule the country," writes Murong Xuecun today. "Ever since they were little, they've been taught to be loyal to the collective, the Party, the country. There's only one thing they're not loyal to: themselves." In a country like this, it makes sense to deliberately steer clear of information. The Chinese are no fools, writes Chang Ping, who today lives in exile in Germany, and was the editor of the *Southern Weekend* when it was still the best newspaper in China. But, he says, many people are consciously choosing not to think. "Because thinking leads to understanding, and understanding only leads to trouble."[162]

I've heard these thoughts expressed often. "The truly unlucky people in this system are the ones who have seen through it," a Beijing teacher told me. "The best thing is to carry on being one of those people who walk through life in a fog of confusion: then you're safe." And the artist Ai Weiwei once wrote on Twitter: "As soon as you try to understand your motherland, you have set off down the path of crime."[163] Very few people have the courage to tread the suicidal path walked by civil rights lawyers and dissidents. For the rest, knowledge only makes living a lie more agonizing.

On a trip to Taiwan with a party of Chinese tourists from Chengdu, I saw for myself the extent of people's determination not to let potentially disturbing information get through to them. One member of the group worked as a Party secretary in a municipal authority. She was a young woman, fashionably dressed, with whom I had lively conversations about Taiwan's night markets and its wonderful food. "We used to have great night markets like that in Chengdu," she said. "Sadly, they were all torn down." We were standing near the Sun Yat-sen memorial hall, waiting for our bus, when we were approached by an older man whom we quickly recognized as a member of the Falun Gong sect.

With its mixture of gentle exercise and eclectic teachings inspired by Buddhism and Taoism, Falun Gong drew millions of followers in the China of the 1990s, until the Communist Party started to feel threatened. The general secretary of the CCP at the time, Jiang Zemin, banned Falun Gong and began persecute its members, sometimes in the most brutal way. In Taiwan, Falun Gong followers now deliberately position themselves in spots that they know Chinese tourists will pass, to inform them about the persecution of their fellow believers on the mainland. They carry placards and brochures bearing gruesome images of torture and organ removal, which they claim are scenes from Chinese prisons and camps.

We were standing on the pavement, then, when the man approached us. He headed for the young Party secretary beside me and thrust his brochures toward her. I saw the surprise in her eyes, which quickly turned to mild panic—but what happened next completely took me aback. First she seemed to freeze, then she put her hands over her ears, squeezed her eyes tight shut, and started to stamp both feet on the ground like a little girl, shouting: "I don't see anything! I don't see anything!"

All in all, it was a remarkable trip. It was the first time any of my tour group from Chengdu had been to Taiwan: the island that has gone its own way since the end of the Chinese civil war in 1949. For a long time, Taiwan was a dictatorship, ruled by the nationalist Kuomintang under Generalissimo Chiang Kai-shek. Then in the mid-1980s, the island took the first steps toward democratization under Chiang's son, Chiang Ching-kuo. To this day, the CCP in Beijing claims Taiwan as a province of China, but over the past three decades the Taiwanese have transformed their home into

Asia's most vibrant democracy, and they have no desire for reunification.

The Chengdu party were a thoughtful group, and after just a couple of days in Taipei they started bombarding the tour guide—and each other—with questions. How come people here are so polite and nice to each other? Why are passers-by so helpful, even though they don't know you? Why is the traffic so orderly, unlike at home? Why does everyone, even the pedestrians, stop at red lights? "Well, we have made our own rules," the tour guide (a native of Taipei) said laconically. "And we just stick to them." My group was genuinely baffled by what they observed from the bus, and some were ashamed. Weren't these people Chinese, too? Didn't they come from the same culture, speak the same language, have the same black hair and the same DNA? And yet the Taiwanese treated each other so differently.

Taiwan is interesting not just because it's a rare example of a successful, non-violent transition from dictatorship to democracy. It is also the living rebuttal of the CCP's claim that the Chinese aren't cut out for democracy. (Officially, Taiwan calls itself "Republic of China" to distinguish itself from the communist "People's Republic" on the mainland.) During Taiwan's transformation, it was fascinating to observe, close-up and in real time as if under a microscope, what the change was doing to society; to see what influence the new political system was having on the way citizens interacted. When I arrived in Taipei as a student in 1987, President Chiang Ching-kuo was on his deathbed, Taiwan was still officially a one-party dictatorship, and Taipei was an Asian city like so many others—exciting, ambitious, chaotic, colorful, and dirty. The air stank, the traffic was always gridlocked, and of course no one ever stopped at red lights. Then came the first free elections, and the Taiwanese threw themselves into the project of democracy with a passion I had never seen in Europe. A society was beginning to stand on its own two feet.

I had an "Aha" moment in March 2004, on the night of Taiwan's second presidential election, which the former opposition leader managed to win. I was there as a foreign correspondent, and at around midnight I left one of the election-night parties where the supporters of the Democratic Progress Party were celebrating their candidate's victory. It was pitch black outside, and hardly any cars were still on the road. I set off for my hotel and arrived at a set of traffic lights at the same time as a group of other people. The pedestrian light was on red. I did what I have done all my life in Asia (and what I'd always used to do in Taiwan): I ignored the light and carried on walking.

I'd taken three paces out into the road when something happened that I'd never experienced before. Someone called out: "Hey!"—then again, louder: "Hello!" I turned around. All the other pedestrians waited dutifully at the lights. And the one who had called out pointed upward. "The light! Red!" There were no cars to be seen, but the Taiwanese had still stopped at the side of the road. I felt like the Chengdu tour group did a few years later: I was momentarily baffled. Then I was terribly ashamed, and trotted back to the pavement like a good citizen.

If you want people to stop at a red light, there are two ways of doing it. You can persuade them with total surveillance, cameras, big data, and the sanctions doled out by a Social Credit System. Or you can give people responsibility for themselves and let a society make the rules it considers necessary for smooth day-to-day coexistence.

I would urge anyone planning a stay in China to add a few months in Taiwan or Hong Kong—especially if they want to discover what makes "the Chinese" tick. Observing how Chinese communities live in a democracy under the rule of law sharpens the eye. You begin to differentiate what might be essentially "Chinese" about the people in mainland China from those aspects of their behavior that result from an interfering political system.

When state power is continually used to manipulate people, the first casualty is the human heart, says the South Korean ex-diplomat and academic Ra Jong-yil, who has written much about North Korea.[164] "When totalitarian control of a society continues for a long time and when there are no alternatives open to the people except to adapt to the external coercion, the result is a kind of Stockholm syndrome on a national scale. People internalize the repressive system to improve their chances of survival, to facilitate accommodating what is inevitable. People not only go along with the system, they even really believe what they have to, what they are taught to." In China, you will often hear people defending the Party's distortions of reality with utter conviction, no matter how implausible they are. These people don't know any better. Except that a lot of the time they do, and are just pretending.

The art of dissimulation becomes second nature when you live under an authoritarian regime. It is true that some people will completely absorb the lies via osmosis and eventually forget the difference between lies and truth—but they will always be in the minority.

Most people wear a mask, which they sometimes keep on even within their own four walls: if their children go to school and repeat something they heard at home, it could get them into trouble. "The state chooses what's mainstream, and you have to conform to that. If your ideals are not mainstream, then you're wrong," the young entrepreneur Wang Sicong, who makes his money from online games, told the BBC: "Why is online gaming becoming so popular in China? Because once you go online, you can take off that mask and say whatever you really think rather than what is mainstream."[165] Wang Sicong is the son of the real estate mogul Wang Jianlin, who for many years was the richest man in China.

Mei you anquangan. "There's no sense of security here." Everyone bobs and weaves; no one keeps still. Is that dynamism? Perhaps. Above all, though, it is a kind of mass nervousness. In today's China, the uncertainty seems even greater than it used to be. In Europe, material gains have brought a feeling of security, at least for those who have joined the middle classes. Not here. People in the cities are vastly wealthier than they used to be. At the same time, the pressure they are under has risen exponentially. Property prices have shot up much faster than salaries. An engineer at the iPhone manufacturer Foxconn in Shenzhen worked out for me, on the tablecloth of a pizzeria a short walk from his factory, how long he would have to work in order to buy a two-room apartment in the neighborhood. It was more than 200 years. He had a degree from the university in Harbin.

So far, there are only the beginnings of a functioning welfare state. "If one person in the family gets cancer, it's enough to drive the whole family into financial ruin," I was told by a friend in Beijing, an ad-man on a good salary. "They call people like me 'middle-class' now, but the term has a very different meaning from the one it has in Europe." Another acquaintance, a PR manager for a large chain of restaurants, agrees: "No one feels secure here, no one, no matter how much money they have. In Germany, you live; here in China, we run for our lives."

The pace of change and the pressure of competition in the new China is one problem. Another is the nepotistic system inherited from the old China, which makes it impossible to navigate life's major challenges—finding a school for a child, or a hospital, drugs, and blood-transfusions for a mother with cancer—according to simple rules that apply to all. At every turn people will have to draw on their network of connections, show that they know the right people, call in favors. People don't have rights they can invoke, or courts that will protect them; nor do politicians feel accountable. After the devastating explosion of the illegal toxic chemical store in Tianjin, right next to a high-rise development inhabited by well-off citizens, a much-shared essay on Weibo contained the following passage:

> Obviously, if you live in a nice house in Binhai with a BMW and a little dog, in your free time you twiddle your fancy worry beads, or else you go for a run and get your exercise in. You maintain a noble silence on any public incident you're aware of. On the surface, you look no different from a middle class person in a normal country.
>
> But this is a delusion. One explosion later, and the homeowners in Qihang Jiayuan and Harbor City discovered they're the same as those petitioners they look down on, making the same moves: kneeling and unfurling banners, going before government officials and saying "We believe in the Party, we believe in the country." This method has been used by countless petitioners—people from the provinces who make five- or six-hundred yuan a month and receive chemical fertilizer subsidies. The homeowners realize, much to their embarrassment, that after an accident there's really #nodifference between us and them.[166]

We are all the underclass. The autocratic state is a minefield for every individual living in it. No matter how much money or power you've amassed, no one is safe. The laws are deliberately made so that everyone breaks them every day. And in the end, you can't rely on them anyway. If the Party has decided it's time, then even the bosses of the biggest companies can vanish overnight. The same fate can befall any Party functionary, no matter how far he has climbed up the ranks. Life at the top of the Party can be "nasty, brutish, and short."[167] Invariably, corruption is the official charge leveled against those who have been arrested, but in reality taking bribes, tax evasion, and smuggling are misdemeanors that almost everyone in China's elite has committed to a greater or lesser degree. They are only used against you when you find yourself on the wrong side in the game of power.

Overnight, Party figures whom the propaganda machine once painted as halo-wearing guardian angels become black-hearted devils. Lu Wei is a good example. For years, he was the head of the Cyber Administration, silencing China's bloggers and setting up the World Internet Conference for Xi Jinping. He was courted by Mark Zuckerberg and in 2015 was named by *Time* magazine as one of the world's 100 most influential people. The day after his arrest in February

2018, the Party press revealed to the astonished public that this same Lu Wei had always been a "shameless" and "tyrannical" person: "He exchanged power for sex. He used his position for personal gain and had numerous properties given to him as gifts. He betrayed every single important political principle. He is a typical hypocrite. His case is extremely bad and particularly serious." Not long afterward, I was invited to the Foreign Ministry on an entirely unrelated matter, where a civil servant revealed to me in passing that he had worked for Lu Wei ten years previously. "Even then," he said, "we all knew what a bad apple he was." If this was true, how come he managed to rise through the ranks for another ten years, all the way to the top of the Party and the nation?

It is the irony of such a system that not even the autocrat himself can be sure of his freedom or his life. Day and night he is plagued by fears of all those rivals he suspects are secretly planning their revenge for the day after he is toppled. In China there is an ancient four-character saying that captures the ruthlessness of such power struggles: *Ni si wo huo*—You die, I live. Perhaps this explains why Xi Jinping has altered the constitution so that he can rule for life. Even for him, nothing less can guarantee a basic level of security.

The Iron House

How a Few Defiant Citizens
Are Refuting the Lies

"The refusal of one decent man outweighs the
acquiescence of the multitude."

Sima Qian, Records of the Grand Historian,
c.100 BC

I magine an iron house. Large, windowless, inde-
structible. A lot of people are asleep inside the house,
not realizing that they will soon suffocate. You know
that certain death awaits them, but you also know they
won't feel any pain. "If you cry out now to wake a few
of the lighter sleepers, and make those unhappy few
aware that they will suffer the agony of an irrevocable
death—do you think you will have done them a good
turn?"

The lines come from Lu Xun. He talks about how he wrestled with his conscience before publishing his first story, *Diary of a Madman,* in April 1918—a milestone for China's emergent modern literature. It is in this story that the term "feast of human flesh" is used to describe China. "There are the lofty and the lowly, great and small, the top and the bottom. You are mistreated, but you can also mistreat others; you are eaten, but you can also eat others." What do you do with this knowledge: call out to the people who are trapped in Lu Xun's iron house? Do you try to free them? Against all reason? "If a few of them wake up, you cannot tell them there is no hope of destroying the house."

Lu Xun decided to write. He wrote and he cried out, trying to wake the sleepers with his words, and in 1936 he died. A few years after his death, a new set of people took over the house; people who had initially entered to wake the sleepers themselves. "Arise! Ye who refuse to be slaves!" the national anthem of the People's Republic of China begins. But since they moved into the house, they have reinforced the iron walls and are now filling in the last remaining cracks.

China doesn't have a long history of intellectual social critics speaking truth to power. Mao had a particular hatred of intellectuals, whom he called the "stinking ninth class." He never forgot the condescension with which they had once treated him, a library assistant and the son of peasants. The gentleman scholars of ancient China had always had a share of power; they looked on the emperor as an equal and practiced a special loyalty. The old books told these scholars that they should speak the truth to a ruler who had gone astray, but the emperors' tyrannical cruelty ensured that most stayed mired in moral cowardice and opportunism.

Sima Qian was the Herodotus of ancient China, the most famous historian the country has ever produced. In 99 BC he experienced at first hand what a single wrong word could cost a person. Sima Qian's crime was to defend a general before the imperial court, calling him a brave man, even though he had capitulated to enemy troops. The emperor had Sima Qian castrated. The scholars, intellectuals, and writers of the millennia that followed learned from examples like this—and generation after generation chose to castrate themselves figuratively. Today, the overwhelming majority still tell the rulers what they want to hear.

"Why are cynicism and shamelessness so widespread in China today?" asked Professor Li Chenjian, vice provost at Peking University, in an essay that was widely shared before being deleted.[168] "Our education system produces genteel, sophisticated liars, not defenders of truth." Li calls on his colleagues not to sell their dignity and independence. The author is not brimming with hope for the future, but nor does he hold the "cowardice and meanness of human nature" alone responsible for the sad state of affairs. In today's China, according to Li, even people's right to remain silent has been taken away: "They are forced to join the jubilant chorus." The essay did the rounds in March 2018. A few weeks previously, Peking University had opened its newest faculty: the "Centre for the Study of Xi Jinping Thought on Socialism with Chinese Characteristics for a New Era." Dozens of other universities had already rushed to set up similar research centers. "What has happened to the glorious tradition of [Peking] University?" the historian Li Ling asked with a sigh.[169] A university should be producing the country's elite. "Those who only know [how] to wait on leaders and bosses are not talents, but slaves."

Is it surprising, then, that a writer like Mo Yan, who won the Nobel Prize for Literature, made his peace with censorship and the apparatus of persecution? Or that he complied without protest when the Party commissioned him to pay tribute to Mao Zedong's infamous 1942 "Yan'an speeches on literature and art" for a celebratory jubilee book? For a fee of 1,000 yuan each, Mo Yan (whose pen name means "Don't speak") and other Party-endorsed authors wrote out Mao's statements, which the Party had used for decades as guidelines for its persecution of artists and writers. In graceful calligraphy, they painted the poisonous words of the Great Chairman, which had once brought torture and death to so many of their colleagues. Nor should it surprise anyone that Mo Yan remained silent on the fate of the other Chinese Nobel laureate, the essayist Liu Xiaobo, whom the state left to waste away in prison until he eventually died of cancer. How many people would have Liu's unbending will and courage in the face of death? If we are surprised by anything, it should be the fact that this country can still produce people like Liu Xiaobo. "It will take a few great men to compensate for the weakness and cowardice of the masses; our country needs moral giants," he wrote in a letter to the poet and author Liao Yiwu, who now lives in exile in Germany.[170] Moral giants are a rare species in any society.

The voices of the other China still exist, even if they can hardly be heard these days. There are the people who protest, the people who cannot help but speak the truth, and the people who refuse to sing with the jubilant chorus. Those who remain decent in the face of indifference and open hostility.

The sinologist Geremie Barmé has been tracking down these voices for decades. "During periods of political darkness such as today," he says, they are "like magma coursing under a hardened mantle."[171] They are the resistance: the writer who adds the redacted sentence back in, even though he knows it will be removed again. The journalist who has been fired, speaking out on Weibo even though he knows his voice will only be heard for a few minutes. The lawyer under pressure from the state security services, who stubbornly persists in taking the law and the constitution at face value. The ordinary citizen who decides to open her mouth for the first time in her life, like Beijing resident Zhao Xiaoli, who was so shocked by Xi Jinping's alteration of Article 79 of the constitution (allowing him the possibility of lifelong rule) that she published an extraordinary essay online.[172] She appealed to all the clever satirists and whispering critics in China to stop camouflaging their anger and their criticism— the time had come to break free from metaphor and be direct. Yes, she was afraid, she wrote, but "because this is intolerable, the furthest extreme of intolerable, I will no longer be silent. I will not satirize or use sarcasm. I will not complain. I will not use metaphor. I will clearly express my point of view . . . In exchange for silence, rulers have repeatedly trampled the power of the people underfoot. For silence, we have the endless lust for power of dictators and autocrats. Silence has brought about the proposed amendment to article

79 of the constitution. For silence, we get the boot from *1984*, stamping our faces forever."

They're just individuals, but the Party still sees them as a substantial threat. "Words still have power. Spoken words are more powerful than secret opinions. Words made public are more powerful than whispered conversation. Explicit opposition is more powerful than metaphor," she writes. An attempt to step outside the lie, even if it is only made by a seemingly weak individual, is an attack on the whole construct of lies. "For the crust presented by the life of lies is made of strange stuff," wrote the Czech dramatist and dissident Václav Havel in his essay *The Power of the Powerless*. "As long as it seals off hermetically the entire society, it appears to be made of stone. But the moment someone breaks through in one place, when one person cries out, 'The emperor is naked!'—when a single person breaks the rules of the game, thus exposing it as a game—everything suddenly appears in another light and the whole crust seems then to be made of a tissue on the point of tearing and disintegrating uncontrollably."[173]

In 1977, Havel and his friends had called the declaration upon which their opposition was based "Charta 77"—and it was no surprise that three decades later, following in Havel's footsteps, Liu Xiaobo drafted a

"Charta 08." No surprise, either, that the Party immediately incarcerated him.

"Actually, all I ever wanted to do was speak the truth. No more and no less. To live in truth, as Havel put it. Isn't that an innate urge for everyone?" the civil rights activist Hu Jia once told me. "But living in truth doesn't just mean being honest with yourself and others; here in China, it also means seeing society and the system clearly for what they are."

The only way I could conduct this interview with Hu Jia was over the phone; it was approaching the anniversary of the Tiananmen Square massacre, and the security services had him under house arrest. When I asked him how many people in this country actually managed to live in truth, he told me about the arrest of his friend, the author Yu Jie, by security service officers at the time of the Arab Spring. "They said to him: 'In the whole of China there are no more than 200 influential critics of the Party like you. If our Party is ever under serious threat, we can take you all away overnight and bury you alive. No one will miss you.'" Hu Jia laughed. "So there are those 200, at least."

Simply telling the truth is what makes people like Liu Xiaobo and Hu Jia a danger to the Party. In a state like China, people who resist are the last reminder that another life is possible. They must therefore not only be silenced; they must also be smeared and vilified as moral or sexual degenerates. That is one purpose of the public shaming sessions on state television, where bloggers and civil rights lawyers are paraded before the cameras after their arrest, forced to denounce each other, and finally "exposed" as greedy fraudsters and sexual predators. The moral giants must be shrunk and turned into squalid gnomes for the comfort of the masses. *They're no better than we are.*

Saying out loud what others can't say, or don't dare to. That's what makes the dissident into a dissident, as Václav Havel wrote. And his "most positive and maximal program" is the simple defense of humans and humanity. There aren't many of them left, and the Party is so dominant that their actions come across as Don Quixote-ish madness. And yet they continue to exist; each generation brings forth its own seekers of truth. And especially in times of crisis, some of them manage to grab the attention of a lot of people for at least a few moments. During the coronavirus crisis, for example, there were intellectuals like Tsinghua Professor Xu Zhangrun, whose essays in the spirit of Lu Xun and Liu Xiaobo don't hesitate to call the emperor naked. And there were people like Chen Qiushi (a lawyer), Li Zehua (a former employee of the state television), or

Fang Bin (a Wuhan citizen and blogger). When the crisis broke out, these three simply took their smartphones and cameras, went to Wuhan's hospitals and crematoria, and filmed everything they saw, then streamed on social media all the things that the authorities didn't want anyone to see. In other countries this is called citizen journalism, while the CCP calls it a threat to state security. Li Zehua's camera was still running when he was kidnapped by agents, and all three of them had disappeared at the time this manuscript was written. Without ever being charged, without trial.

"What we do know for sure is that this machine, increasingly, can do whatever it wants. The monster is fully grown," wrote Xiao Meili, a Beijing women's rights activist born in 1989, on her WeChat account in 2018, referring to the security apparatus:

There are people who left comments to ask me: "Why didn't they go harass other people? Why did they only harass you? What is the matter with you?" And I understand: when people are living in an environment where resistance isn't allowed, they can easily long to believe in the powerful. Otherwise how do they deal with their situation? It takes courage to accept the truths of the world that one lives in.

From time to time, some people would say to me: "You all shouldn't be that radical. Don't provoke them. Only then can we get things done." But you'd be radical, too, if you didn't have all these other people in your way[. . .]

I asked my friend Lu Pin, "What should we do in this ever more terrible environment?" Lu Pin said: "Live on. Outlive them. Only then can we see hope." Yes, we need to live on with healthy minds and bodies. And I hope we can grow too. I want to use their words to document our flesh and bones. In a cyberspace full of hostility, I hope to carve out enough room for a hug, and build some trust between people that cannot be easily broken.

This is why I wrote down these stories.[174]

The courage for this kind of self-sacrifice and heroism is only possessed by a few. But the truth can survive on much less. Writer Yan Lianke has one hope for the post-coronavirus era: "If we can't be a whistle-blower like Li Wenliang, then let us at least be someone who hears that whistle. If we can't speak out loudly, then let us be whisperers. If we can't be whisperers, then let us be silent people who have memories. Having experienced the start, onslaught, and spread of Covid-19, let us be the people who silently step aside when the crowd unites to sing a victory song after the battle is won—the people who have graves in their hearts, with memories etched in them; the people who remember and can someday pass on these memories to our future generations."

Perhaps Mao Zedong was right, and a spark really is all it takes to set the whole steppe alight. In a bed of cold ashes, the brief glimmer of these sparks stops the last hope from being extinguished. And there are still the children, as Lu Xun wrote in 1918, who have not yet eaten human flesh. "Save them, save the children!"

The Gamble

When Power Stands
in Its Own Way

"Obey the Party!"

*Inscription on mooncakes for the Mid-Autumn
Festival in September 2015*

Some of the Party's plans read like science fiction. But it isn't the first time communism has dreamed of a digital rebirth. Soviet computer scientists were doing that more than half a century ago in Kiev and Moscow, having discovered cybernetics.[175] They, too, worked on high-tech fantasies designed to replace ideologically undesirable phenomena (at that time, the free market economy) in the Soviet system, which was heading for crisis.

The cyberneticist Victor Glushkov presented his

"National Automatized System of Administration of Economy" (OGAS) in 1962, with the aim of achieving "electronic socialism." Glushkov wanted to set up a nationwide computer network, linked by the country's phone lines to every factory and company in the command economy. It would be the smart nerve system of the Soviet economy, giving feedback in real time, enabling rational decision-making processes and even the invention of an electronic currency. In the end, Glushkov and his colleagues failed very prosaically, when their plan was crushed by scheming ministers and bureaucrats.

This time, the launch conditions are better. Crippling bureaucracy and scheming ministers exist in China too, of course, but the tailwind blowing from on high is strong. Even so, Xi Jinping has taken a huge gamble. Will he manage not only to reinvent dictatorship, but take it to the heights he is currently dreaming of? In 2049, its 100th anniversary, will the People's Republic of China be that "modern, prosperous, strong, democratic, civilized, harmonious, and beautiful country" about which Xi Jinping gushes to his people and to the world at large? Will it be under the leadership of an "even stronger" Party, ruling over all known points of the compass and those yet to be discovered? A country "at the center of the world," with a model of strong leadership and a booming economy that will become "a new alternative for other nations," and boasting a "world class" army, "ready for battle"? Will it work? Or is the system itself becoming its own worst enemy? There are indications that the latter may be true.

Will there still be a People's Republic in 2049? Or a Party? Xi Jinping would be 96—not a wholly unlikely age, given the general longevity of the red aristocracy. Xi's father Xi Zhongxun died at 89; Mao Zedong lived until 82; and Deng Xiaoping until 92.

The Party has past experience of reincarnation. This is something that sets it apart from all its communist brother-parties. It's why the CCP is still alive and ruling, while the others have passed away or been twisted into caricatures of their former selves. Its first two rebirths—shortly after Mao's death in 1976 and then again, to a lesser extent, following the Tiananmen Square massacre in 1989—were extraordinarily successful, against all the odds. The Party showed just what it could do: overcome the laws of nature, defy gravity. It could import capitalism and at the same time remain true, if not to the idealist Marx, then certainly to the power-hungry Lenin.

That was the service Deng Xiaoping performed. The Chinese people embraced globalization and created an economic boom, with miraculous growth rates that seemed even more miraculous because of the destruction that the Chinese Communist Party had wreaked upon the country, leaving a very meager economic base from which to start.

Crucial to Deng's achievement was that he kept politics off the people's back to an unprecedented degree. He condemned the Maoist cult of the personality and ruled that the Party should be led by a collective. He even started to talk about the separation of Party and state, at least in the years before 1989; and he granted new freedoms to society outside the realm of economics. Hence the much-vaunted pragmatism of China's Communist Party, its legendary ability to change and adapt. And hence the boom that Xi Jinping inherited, upon which he is building his dream of the "rebirth of the great Chinese nation."

Today Xi is breaking with all of this. He is bringing back ideology, drawing all power to himself, suffocating experimentation. Once again he is sealing China off from the world; once again he is placing the all-seeing, all-knowing Party at the heart of things.

But Xi Jinping is no Mao Zedong, and he certainly isn't just another central Asian kleptocrat. He really does have a vision for China. Many Chinese people admire him as a strong man who fights corruption and is making their country into a proud nation. The mix of nationalism, siege mentality, and dreams of superpower status pumped into the people by the propaganda machine is highly effective. As long as the system avoids a major crisis, Xi—like other autocrats before him—will no doubt feel confident that his freedom- and information-starved people are behind him.

Nevertheless, at the start of Xi's second term in office, there were signs that absolute power has its own self-defeating logic, driving Xi Jinping and the Party to do themselves and their plans more harm than good. Before Xi, even the more liberal sections of the middle classes were prepared to accept the Party's rule, as long as it provided some of the advantages enjoyed by open societies: increasing wealth, a functioning administration, the expansion of modern infrastructure, growing freedom in their private lives.

In China the social contract has always been unwritten, but—at least for the urban population—that didn't make it any less real. Hope and optimism have loomed large in the emotional lives of those growing up in China over the past 40 years. Things were getting a little better every year, and that made people believe in the future. This eternal hope also helped to ensure that there was very little rebellion over the environmental devastation, the obvious inequality, and the shameless corruption of a greedy elite. For a long time, however ignored or disadvantaged you felt, you could take solace in the thought that tomorrow would be your day. Or at least your children's day.

The belief in a better future—better partly because it would be freer—has remained a constant. "We look at each new leader like the sickle of the new moon," a liberal writer said on the occasion of the Mid-Autumn Festival in 2012, shortly before Xi Jinping took office. "We are filled with confidence that this time he will grow round and full. And we never give up hope, even when one leader after another disappoints us."

Three years into Xi's rule, mooncakes suddenly appeared at the Mid-Autumn Festival. They were being handed out by the police, and had the characters "Obey the Party" stamped on them. And three years after that, Xi Jinping had the constitution changed to allow him to remain president for life. The number of searches on Baidu containing the term "emigrate" shot up. Five years had passed since Xi started dismantling the freedoms of Chinese citizens, piece by piece—but it was this move that shattered something for many people. Now their hope had gone, and they finally stopped believing in the Party's pragmatism, its will to reform. Yes, they had been living under a dictatorship, but at least it had been trying to institutionalize its rules and procedures, trying to make its exercise of power more

predictable, trying to take account of conflicting interests. When the Party's willingness to experiment, to
be flexible and adaptable, became a thing of the past,
the pillars of political stability began to crumble. Yes,
the middle class has been the great ally of the CCP
for the past three decades. But it was because the party
delivered more than hollow phrases: reputation in the
world, modern infrastructure, and above all material
prosperity. But what if the latter suddenly fails to ma-
terialize, if the economy slows down or even collapses,
if the party can no longer deliver?

Xi Jinping brought back consistency—but it is a
merciless sort of consistency. The China of recent decades had been a country of *chabuduo*. Roughly translated, *chabuduo* means "more or less"; something like
"cutting corners." *Chabuduo* meant that laws often
weren't enforced 100 percent; that the censors didn't
always look at things too closely; that the authorities
frequently turned a blind eye. It was the Party's way
of giving breathing space to a society that was increasingly self-aware and sophisticated. Chinese life outside
the sphere of politics could sometimes seem relaxed to
the point of anarchy, and for many people, it fed the
illusion of freedom.

Xi Jinping isn't giving anyone room to breathe.
But rigid systems become brittle, and absolute power

not only corrupts—it can also obscure. In 2017, an academic journal in Beijing published a paper entitled "The Application of Marxism in the Analysis of Ozone Levels in Beijing." The paper is not (yet) representative of scholarship in China as a whole, but it is a sign of the times all the same: the last time academics bent over backwards to produce this kind of nonsense was under Mao.

Many disillusioned Chinese citizens will try to emigrate. In terms of numbers, this won't cause the Party any pain—and after all, it will be easier to preserve stability if dissatisfaction goes elsewhere. But what if emigration starts to take away the brightest and the best? Xi's speeches have long insisted that "innovation" and "creativity" are indispensable to China's future. Among the frustrated and disappointed emigrés will be many innovative, creative minds—in a closed system like Xi's, they are the first to come up against limitations. Such a system soon stifles the flow of energy and ideas that it has recognized in theory as central to its plans. This is an even greater risk in a country with a rapidly aging population: at the start of 2019, China's National Office for Statistics counted more people over age 60 (249.5 million) than under 15 (248.6 million) for the first time.

Xi's one-party state is strong, and the Party leader

is doing everything he can to make it even stronger. The Party has the tools of repression and the financial means to nip any political alternative to its rule in the bud. For the most part, these things ensure the loyalty of the elites and the support—or at least the resigned indifference—of the masses. In public, Xi exudes tremendous self-confidence, but the level of repression, the unprecedented rise in surveillance, and the strengthening of the security apparatus tell a different story. The Party knows that the acquiescence of the people is a wavering thing. "The ruler is the boat, and the people are the water," wrote the influential Confucian philosopher Xunzi in the third century BC. "The water can bear the boat up, and the water can capsize it." The Party is well aware that the support of citizens rests on increasing prosperity first and foremost, but also on information control, mind control, and the suppression of dissent. If there is one lesson to be learned from the events in Tiananmen Square in 1989, from the downfall of communism in Eastern Europe, and from the collapse of the Soviet Union, it is that the boat can tip quickly—the instant that rulers show signs of weakness. The current of resentment that is always beneath the surface will then become a force to be reckoned with.

Dangers lurk along the path that Xi Jinping is now

taking. The future is likely to hold increasing repression, and an even more elaborate personality cult. But when a leader ceases to tolerate any opposition, even from loyal friends, leaving himself surrounded by flunkies and yes-men, he makes the system susceptible to mistakes that, with no mechanism to correct them, can have serious consequences. If one forces an entire people to lie, in the end Xi's last advisors left will lie too. The silence of the intellectuals and academics, the many civil servants paralyzed with fear, means that important feedback from the provinces and counties, from social and economic interest groups, only reaches the Party leadership slowly and in a distorted form—or else not at all. The AI evangelists theorize that algorithms will fill this gap, beaming any signs of an approaching crisis all the way up to Xi's headquarters. In practice, though, AI machines are only as good as the input they receive: if algorithms are fed rubbish at the bottom, they will spit out rubbish at the top. Local functionaries and executives in China are past masters at data manipulation.

The coronavirus crisis revealed just how much Xi Jinping's addiction to control has already weakened the immune system of the society: When the SARS virus hit China in 2003, it was also a huge crisis—but at that time there were still activists around and investigative journalists in some of the more courageous state-controlled media, independent voices that could alarm society and the bureaucracy. But Xi Jinping has wiped out civil society, he has castrated the traditional media and silenced the internet. Society no longer has an early warning system. But it's also the apparatus of power that blinds itself. "When the media is surnamed 'Party' . . . the people are abandoned," wrote Party member and entrepreneur Ren Zhiqiang in his essay on the coronavirus crisis. The problem is the system amputating itself in its obsession with power: "Even if we resolve this epidemic, there'll always be another disaster down the road." Beijing law professor Xu Zhangrun argued along similar lines in his essay. Xi, he said, has gathered a sycophantic court around him, and with his primacy of ideology over competence and professionalism he has created a "system of impotence": "Although everyone looks to The One for the nod of approval, The One himself is clueless." A China ruled by such a system, Xu said, could never be more than a "crippled giant."

The coronavirus showed the whole world, but above all its own citizens, how the Chinese system reaches its limits dangerously fast in times of crisis. Chinese scientists did an excellent job and were able to quickly identify the DNA sequence of the virus. In the greatest possible contrast to this world-class research was the failure of the system, which is still hostage to anachronistic political reflexes from the early 20th century. In a sense, the Party is now also victim of its own successes: its policies have created the conditions for such cutting-edge research, just as its economic strategy has fostered the emergence of what is today the largest middle class on earth. But while this well-educated and by now well-traveled middle class may be silenced astonishingly effectively in times of a flourishing economy and apparent normality, they feel betrayed the moment they have to fear for life and limb because of the breakdown of the political system, the moment they observe how the party apparatus spends more energy on targeting outspoken doctors than on containing a deadly virus. When it takes its logic of repression too far, the system often trips itself up. That could be seen twenty years ago, with the example of the Falun Gong. In 1999, the then Party head Jiang Zemin launched a campaign of merciless persecution against the sect. The result was that, in the space of a few years, what had been a well-organized religious group inside China whose followers were mostly of retirement age became a network of noisy activists spread across the world. Today, Falun Gong runs a global media empire with a single primary aim: to topple the Communist Party.

Like his predecessors, Xi Jinping prefers *ad hoc* campaigns and popular mobilization to stable policies and laws. As a result officials across the country vie with each other in their eagerness to divine the ruler's will and get into his good books. If Xi wants to clean up the air in Beijing, overhasty cadres will remove the coal-stoves from millions of Beijing residents' homes in the middle of a bitterly cold winter. If Xi wants to see the capital "civilized," then overnight they will throw tens of thousands of migrant workers out of their apartments and tear down the roofs over their heads. This zealotry is not only inhuman, it also poses a threat to those in power.

More and more, you get the feeling that the system is needlessly making enemies. When it locks up feminists protesting against sexual harassment. Or when the Party decides to take on humor—by banning puns, for instance. This happened in 2014, when the State Administration of Press, Publication, Radio, Film, and Television in all seriousness issued a ban on wordplay, declaring that the playful alteration of old sayings led to "cultural and linguistic chaos." It was a particularly absurd move because Chinese, with its many ambiguous homophones, is better suited to wordplay than almost any other language. For thousands of years the Chinese people have played with their language, with a relish seldom encountered elsewhere in the world.

In 2018, the radio and television authority ordered a company called Bytedance to withdraw its popular humor app Neihuan Duanzi; at the time 17 million users were exchanging funny images and videos, sketches and jokes. The censors called the app "vulgar" and "inappropriate." They had already conducted a campaign against online celebrity gossip, and forbidden not just the country's rappers but also its professional soccer players from showing off their tattoos. In early 2019, they took the legendarily popular period drama *Story of Yanxi Palace* (300 million viewers) off air, after Party organizations complained that all the intrigues and luxury of the imperial palace were corrupting the good souls of the thrifty and hard-working Chinese. Under Xi Jinping, the state apparatus can come across as having an almost Calvinist hostility toward fun, a zealous rigor that is as much moral as political—and which was unknown in previous decades. It has put noses out of joint, not just because people are being denied harmless pleasures, but also because the Party's double standards are so plain to see.

Every power carries within it the seed of its own destruction, and China's past leaders were well aware of this. It's dangerous for any regime to become so intoxicated with its own power that it loses sight of this fact. If the CCP ceases to tolerate fun, it will also have ceased to understand its people—with or without artificial intelligence to help. And if it ceases to understand its people, the model that has served it so well in recent years—the subjects focus on getting rich, buying things, and having fun, while submitting to surveillance of their own accord—is under threat.

For the time being, though, the rule of the CCP is in no danger. There is no reason to doubt the Party's ability to create the most perfect surveillance state the world has ever seen over the next few years. But the question remains: in the end, will this be a state capable of overtaking the West and sprinting ahead to lead the world?

The Illusion

How Everyone Imagines
Their Own China

"Whoever talks about China talks about himself."

Simon Leys (1935–2014)

The Belgian sinologist and author Simon Leys was one of the cleverest and most clear-sighted observers of China. In the 1970s he found himself up against a particularly deluded type of China fantasist: the European Maoist.

For centuries, ever since Marco Polo first sparked mercantilist fantasies of the Far East with his writings about distant Cathay, Europe's seekers of meaning and profit have been gripped by an overblown enthusiasm for the country. "Who would have believed," wrote the philosopher Gottfried Leibniz in 1697, "that there was a people on this earth who [. . .] still surpass us in comprehending the precepts of civil life?" For us, China has always been a blank sheet of paper: virgin territory for our projections. Sometimes China was heaven; sometimes hell. The Yellow Peril. Rarely have there been shades of gray. The China Show is at its most rewarding when you don't even watch: then you can usually find exactly what you were looking for."[176]

In recent decades the West has had a particularly distorted image of China. Under Deng Xiaoping, a brand new type of autocratic state emerged—one that blended Leninist repression with a turbo-charged super-capitalism and the gleam of commerce and consumption. The world had never seen anything like it, which meant many people refused to believe it could exist. It had to be merely a transitional step on a path of transformation. *In the end, China will become like us,* they whispered.

China confuses people. More than a few seek comfort in the popular illusion that economic growth will automatically make the country more democratic. *Wandel durch Handel*—"change through trade"— as the catchy German phrase has it. Proponents of this theory have included European leaders, the International Olympic Committee, and most foreign businesspeople visiting Beijing. Who knows how many really believed it, and how many merely found it expedient. In any case, right from the start the argument had a serious flaw. It wasn't true.

"China's success story is [. . .] the most serious challenge that liberal democracy has faced since fascism in the 1930s," wrote the China expert Ian Buruma more than a decade ago.[177] Communism might be dead in China, but the Party's rule was still very much alive. Yes, the color of power had changed. A regime with a quasi-religious leader-cult, outbreaks of ideological ecstasy, and parade-ground drills had become a regime in which the old Party elite and a new business class joined forces to plunder the nation's wealth, and bought the goodwill of the urban middle classes by offering them a share in the prosperity.

China's leaders have always been adept at pulling the wool over outsiders' eyes—and they are only too happy to be deceived. Spectacular economic growth and a successful drive to combat hunger have made their task even easier. Didn't the residents of Beijing and Shanghai now drive Audis? Didn't they sit in Starbucks drinking cappuccinos, watch MTV, and go on city trips to Bangkok, Paris, and London? Westerners looked for what was familiar—and found the Chinese caught up in a frenzy of consumerism. These people were just like us, after all!

Among China correspondents there has long been a running joke about the columnist from Washington, the banker from London, or the CEO from Frankfurt, who flies into Beijing or Shanghai. After a day, he has formed a lasting opinion on China; after a week, he gives the world the benefit of that opinion in a guest article for the *New York Times*, the *Financial Times*, or the *Handelsblatt*; and after a month, he has enough material for a book. This character is still very much alive and well; like so many other clichés, he simply refuses to die out.

"China Is Leaving Donald Trump's America Behind" is the headline above one of these opinion pieces in the *Financial Times*, by the venture capitalist Michael Moritz. The article begins with a striking claim: "A week in China is enough to persuade anyone that the world has spun back to front."[178] The final paragraph is even better: Moritz urges Donald Trump to "send the managers of his pocket-sized portfolio of hotels to visit the best hotels in Beijing and Shanghai." There, they would find "a level of service unparalleled in New York, London, or Paris." The view of China from the windows of luxury hotels is always a particularly pleasing one.

More than a few Westerners fall for the Party's line that China is unique among the world's nations. Those with a certain level of prominence are flattered and cultivated as "friends of the Chinese people." They might be given a consultancy post in one of the state's countless decorative committees, for example, where they will talk about the "open and honest communication" they enjoy with their Chinese counterparts. At the World Internet Conference in Wuzhen, I met Werner Zorn, the man who first brought the internet to China when he was a computer scientist at the University of Karlsruhe. As a gesture of thanks, the retired professor still gets to sit on a "High Level Advisory Council" and is invited to China twice a year. According to Zorn, Xi Jinping was "a stroke of luck for the Chinese." The action he was taking to combat corruption was unprecedented. In any case, said Zorn, "Just look at the revelations at home about offshore companies, in the Paradise Papers. We're no better than they are." As for the Party's tightening grip on China—well, it was "a large state holding" after all. The Western system would be no good for China: "An opposition here who said, 'No, you can't do that,'"—the professor shook his head—"No. They need a wise leader, and you can certainly call Xi Jinping that."

One hears the fairy tale of the wise dictator somewhat less often these days, but the myth of China as an exotic land that can't be grasped using Western logic is alive and kicking. It's nonsense, of course, if only on the grounds that contemporary China has taken as much from the West—communism, capitalism, entertainment, music, clothes, city planning, science, and technology—as it has from ancient China. And much of what still sets us apart today—family and clan structures, say—is less due to innate differences than to the fact that, until very recently, China was an agrarian society. Anyone who parrots the CCP mantra that China's "national characteristics" mean that it requires exactly the kind of dictatorship the Party has given it has fallen for a form of orientalism devised by Chinese communists. The cultural relativism that says the Chinese aren't ready for democracy, and don't share our desire for human rights, smacks of Western racism. The remedy is a trip to Taiwan to see Asia's most vibrant democracy in action: the people there are Chinese, too.

Surprisingly, many Western observers are reluctant to describe China as a "dictatorship"—though the Party proudly uses the term itself. Or else they might buy the attractive—and false—idea of the hyper-efficient "developmental dictatorship," which is superior to our own system. Not infrequently it's in their own political or economic interests to do so. Michael Moritz, for example, went on to produce a second essay, in which he urged Silicon Valley to give up its lazy ways and start emulating China's predatory capitalism.[179] Moritz—a venture capitalist, remember—reports with great excitement that in China, people work seven days a week for start-ups, from eight in the morning until ten at night. In their breaks they re-use their teabags, and if ever they want to lay eyes on their spouses, they invite them along on business trips.

Self-interest, though, is not the whole explanation; naivete is also to blame. It's part of the reason why Western politicians and businesspeople spent so long clinging to the belief that China's leaders had the best intentions, for decades taking their fine words at face value. And when there were no actions to follow the words, they gave them the benefit of the doubt and blamed deficiencies in the state apparatus charged with making the plans a reality. Or they simply forgot all about them, instead bathing in the endless stream of fresh promises gushing out of Beijing and into the world.

Sometimes their contortions have been painful to watch. Step forward, David Cameron, who had incurred the wrath of the Party for meeting with the Dalai Lama on his May 2013 visit. When he was permitted to return with a trade delegation in December of the same year, he gave a performance described by one European diplomat sarcastically as fitting the legal definition of prostitution. Cameron spoke of Britain and China's "deep complementary economics," of his desire to "help deliver the Chinese dream," and of Britain's unique advantages as a broker between China and the EU. In return, the Chinese premier was at pains to mention Cameron's "respect for China's core national interests and major concerns." Or to put it another way—well done for not mentioning Tibet or human rights this time.

Cameron's trip was the target of not a little mockery from sections of the domestic press: his co-travelers from UK companies included such global heavyweights as the Cambridge Satchel Company and Westaway Sausages of Devon. Regardless of how those particular enterprises are faring almost six years later, it must be noted that the former British PM is now leading a UK China investment fund heavily engaged in Xi Jinping's "Belt and Road Initiative." So some things worked out well. For some.

Yet Cameron is not alone. Beijing has always made skillful use of the China-blindness of the West, and with the election of Donald Trump as US president, Xi Jinping has been gifted an unprecedented opportunity. Trump's alarming isolationism and revolting conduct are in marked contrast with Xi's well-mannered appearances on the world stage—the smooth speeches where he presents himself as the savior of globalization—and make it easy for Europeans to applaud. It is remarkable. The same people who—rightly—pore over every syllable uttered by Donald Trump and shout "Lies!" are prepared to take Xi on trust when he preaches free trade at the World Economic Forum in Davos; when he invokes the international order; or when he promises that China will open up, and sings the praises of a networked globe.

In reality, he is a far greater protectionist than Trump.

In reality, he is subverting the existing international order on many fronts.

In reality, intellectually and ideologically he is sealing his country off.

In reality, he is severing China's last remaining internet connections to the outside world.

Xi is no Maoist, but he has picked up a few tricks from Mao. His recipe for guerrilla warfare, for instance: *The enemy advances, we retreat; the enemy encamps, we harass; the enemy tires, we attack; the enemy retreats, we pursue.*

The Europeans are starting to wake up. In a strategy paper in March 2019, the European Commission for the first time called China a "*systemic rival* promoting alternative models of *governance.*"180 Marking China as a "systemic rival" was a break with previous partnership rhetoric and a premiere that surprised many. In reality, it was nothing more than the recognition of reality. Especially within the German government, many have for a while now already seen China as the challenge it is. And in the United States, beyond the trade war with China initiated by the Trump administration, a broad front has formed in all political camps: China is seen now as a strategic and systemic challenge.

However, there is a realization that our interests are strong, but our instruments are weak, especially as long as Europe is not united and the US government continues to torpedo the alliance between Europeans and the United States. Crucial years have been wasted because we were wrong about China—and they were years when our influence on the path the country was taking might have been greater than it is today.

Today, China is influencing us.

The World

How China Exerts Its Influence

"Xi Jinping Thought [. . .] has the potential to correct
and transform the existing world order."

China Daily, *January 30, 2018*

In London, the Royal Court Theatre removes a play about Tibetan exiles from its program. For fear of China, says Abhishek Majumdar, the play's Indian author. The British Council has written to the theater, advising that the play would coincide with "significant political meetings" in China and might jeopardize the Royal Court's ability to run projects there in the future.[181]

Or Stuttgart, where the German car manufacturer Daimler quotes the Dalai Lama on its Instagram channel: "Look at a situation from all angles, and you will become more open." A harmless meme, a mere calendar motto. But there is an outcry from China: the leadership there regards the Dalai Lama as a "wolf in monk's clothing." At which Daimler gives the Dalai Lama the boot, prostrates itself at Beijing's feet, and suddenly starts speaking in tongues in its public statements. Daimler regrets its "extreme error," the company insists. They know they have "hurt the feelings of the Chinese . . . Taking this as a warning, we are immediately introducing concrete measures to deepen our understanding of Chinese culture and values." To the public, it sounded as if Daimler had adopted the language of the CCP propaganda, word for word.

Or Ankara, where the Turkish foreign minister promises "to introduce measures to eradicate all media reports criticizing China."

Or Cambridge, where in 2017—at the request of the Chinese censors—the venerable Cambridge University Press removes from its websites some 300 articles containing criticism of China. CUP only reverses its decision following a huge outcry from academics all over the world.

Wait a moment: isn't China supposed to be becoming more like the West? These days the Chinese eat at McDonald's; they listen to Adele and Lady Gaga; they drive VWs, Audis, and Mercedes. Isn't this backward? Are we in the West now supposed to change if we want to keep trading with China? In late 2017 Apple's CEO Tim Cook was still giving voice to the old hope: "Your choice is, do you participate or do you stand on the sideline and yell at how things should be? My own view very strongly is you show up and you participate, you get in the arena, because nothing ever changes from the sideline."[182]

Cook's defense of his company's relationship with China came after weeks of harsh criticism: Apple had bowed to pressure from Beijing and removed undesirable apps from its store that allowed users to bypass the censors. A few months later, Apple signed over its Chinese iCloud server, along with all its customer data, to a firm owned by the government of Guizhou Province—in other words, to the Chinese state. In the second half of 2018 alone, Apple removed 517 apps from its app store at the request of the Chinese authorities, according to its own transparency report. The company was particularly criticized in 2019 when it removed the HKmap.live app during the Hong Kong protests—an app that Hong Kong residents could use in real time to search for and mark focal points of the street protests, including police and tear gas operations.

Apple also deleted the emoji that showed the Taiwanese flag for Hong Kong users. Incidentally, the group now sells every third iPhone in China.

Microsoft worked with the Chinese Military–led National University of Defense Technology (NDUT) in Changsha on research papers on, among other things, artificial intelligence and facial recognition.[183] And Google, the only Silicon Valley firm to have rejected censorship and withdrawn from China in 2010, returned in 2018. It now runs an artificial intelligence laboratory there. "I believe AI and its benefits have no borders," said Beijing-born Fei-Fei Li, Google's head of AI at the time of the move, which did not attract particular attention in the mainstream media.[184]

The company only made headlines when newspapers revealed that Google had secretly been working on a search engine for the Chinese market—a search engine that would provide China's censors and surveillance state with superior Google technology. Internally, it was called "Project Dragonfly"—given what we know about the multiple, all-seeing eyes of these insects, Google engineers either possess a highly-developed sense of irony, or none whatsoever. Following angry protests from its own employees, and from American politicians, Google eventually stepped back from the project.[185] *The Intercept* also revealed that

Google, IBM, and the US chip manufacturer Xilinx had been working on a new generation of microprocessors with the Chinese company Semptian through the OpenPower Foundation: Semptian then used the research results to perfect their internet monitoring and censorship technology.

Oh, and by the way, you haven't heard Lady Gaga in China anymore—not since she chatted to the Dalai Lama about yoga.

So is the West changing China, or the other way around? The country has long been an economic superpower, though for decades it was a fairly circumspect one. Those days are gone. At the Party Congress in October 2017, Xi Jinping proclaimed a "new era." He promised that China would now step "into the center of the world stage." Not content with shaping the Chinese state, China also wanted to reshape globalization according to its own design. For the first time since Mao Zedong, China's autocracy was advertising itself to the world as a role model, at a moment when, according to the official news agency, Xinhua, "Western liberal democracy is sinking into crisis and chaos." Chinese propaganda praises the "Chinese path." And China's leader is making it the envy of all other nations once more: the whole world, says the *People's Daily*, "feels the warmth" of Xi Jinping's "community for a common future of humanity." The propaganda machine isn't just glorifying Xi Jinping Thought as the "new communism" at home—it is now also offering a "China solution to the world"—in Xi Jinping's words: "A new option for developing countries to realize modernity."[186]

The competition between systems is back. Xinhua again: "After several hundred years, the Western model is showing its age. It is high time for profound reflection on the ills of a doddering democracy which has precipitated so many of the world's ills and solved so few."[187] Xi Jinping has given the order "to strengthen China's soft power." The world needs to hear the "China story" at long last. Yes, the Chinese lion was now waking up, said Xi on a state visit to France, alluding to Napoleon's reported *bon mot* about China being a sleeping lion that shouldn't be woken. "But it is peaceful, pleasant, and civilized."[188]

China knew that the wider world was a little suspicious—especially the countries in its immediate neighborhood. But there was "no 'conquest gene' in the DNA of the Chinese people," as Xi once said—and ever since he said it, claims of a Chinese "peace gene" have featured in statements by the country's diplomats.[189] What the propaganda doesn't explain: how exactly the Zhou dynasty's few small territories around the middle and lower reaches of the Yellow River three millennia ago became the vast nation we today know as China. Certainly Tibetans and Uighurs very well saw the Chinese as invaders, as did the Vietnamese when Chinese troops arrived in their territory in 1979.

The CCP is now investigating how to exert its influence on a worldwide scale: on business and political elites; on universities, think tanks, and the media. It is meeting with surprising—and so often unnoticed—success. In Europe it has attracted only brief flashes of attention, as in the case of the conservative British author and civil rights activist Benedict Rogers, who campaigns against the disappearance of freedom in Hong Kong. His mother in Dorset and neighbors in London have received slanderous letters as a result of his activities, in which Rogers is vilified as an enemy of the Chinese nation. In Germany, there was some coverage when the intelligence service warned that the Chinese secret service is using social networks like LinkedIn "on a large scale to extract information and find intelligence sources." The service also reported "a broad-based attempt to infiltrate parliaments, ministries, and agencies." In summer 2018, the *Süddeutsche Zeitung* revealed evidence that suggested China's spies had come very close to recruiting a member of the Bundestag, the German parliament.

In October 2019, the NBA's commitment to China

made headlines. In Hong Kong, people who felt Beijing's stranglehold on the freedom of their city had been demonstrating for several months. And in Tokyo, where his team was playing, Daryl Morey, the general manager of the Houston Rockets, posted this message on Twitter: "Fight for Freedom. Stand with Hong Kong." The angry outcry of Chinese propaganda regarding the two short sentences was astonishing. China is a billion-dollar market for the NBA. Indeed, the controversy flared up shortly after Tencent closed a $1.5 billion deal with the league that guaranteed Tencent NBA game-streaming rights for five years. The Chinese Basketball Association, led by former Houston Rockets star Yao Ming, immediately cut ties with his former team, Chinese state television and Tencent threatened to stop broadcasting the games, and Chinese advertising partners announced a boycott.

The first reactions from the NBA, as well as from other Houston Rockets officials and players, were visibly driven by the fear of losing that billion-dollar Chinese market. "We apologize. We love China," said Houston Rockets star James Harden, still in Tokyo. And the NBA initially distanced itself from Morey, called his tweet "regrettable," and expressed its "great respect for the history and culture of the Chinese people." The NBA's apology, in turn, fueled a debate in the United States about freedom of expression and democratic values in the face of repression, in which commentators and politicians from all camps berated the NBA's buckling as shameful. Then–presidential candidate Beto O'Rourke called it the "blatant prioritization of profits over human rights."[190] NBA commissioner Adam Silver reacted four days after Morey's tweet with a U-turn: on behalf of the league, he now defended democratic values such as freedom of speech. He also said no to the request of the Chinese government that Morey be fired for his tweet.

The controversy sparked by Morey's tweet illustrates the dilemma facing Western companies seeking access to the huge Chinese market in the face of growing repression, and the resulting possible risk to their reputation, especially when, like the NBA, they are in the public eye. Coincidentally the incident also revealed other examples of China's influence. There were news reports that the management of the sports channel ESPN had warned its anchors, commentators, and reporters to only discuss the sports-related side of the controversy and stay clear of the political background. (In 2016, ESPN had signed a strategic deal with Tencent to distribute its programs on the Chinese internet company's digital channels.) Observers were even more astonished when ESPN began to show a map of the world as concocted by Chinese propaganda that identified *de facto* independent Taiwan and almost the entire South China Sea, contested by many neighboring countries, as territory of the People's Republic of China. Unsurprisingly, both ESPN and its parent company Disney have substantial business interests in China.

"The magnitude of the loss will be in the hundreds of millions of dollars," NBA commissioner Adam Silver told a press conference on the sidelines of the All-Star Weekend in Chicago in February 2020. "Probably less than $400 million, maybe even less than that," he said. By that time China's state broadcaster, CCTV, had not yet resumed its coverage. Still, there were already signs of reconciliation. Most notably the NBA stating publicly that they were offering assistance and donating money to the Chinese people facing the threat of the coronavirus at that time, and Chinese internet company Tencent having restarted its coverage of NBA games, showing up to three games a night.

Just as the USA once worked to make the world a safe place for democracy, today's China is working to make it a safe place for autocracy. Unlike Russian attempts to influence the West, which are often aimed at a general destabilization and merely destructive, the Chinese leadership tends to focus on specific interests. At the same time, it is working on a broader strategy to build a network of pro-China opinion leaders and decision makers. The primary aim is to eradicate positions with which the Party doesn't agree—for example, by threatening hotel chains and airlines that dare to refer to Taiwan on their websites as a country in its own right. The companies almost always cave in. Givenchy apologized to China in 2019 for a T-shirt on which Hong Kong and Taiwan were not explicitly assigned to the PRC: "Givenchy always respects China's sovereignty," the company's statement said. Dior made a similar public apology after one of its employees used a map during a presentation that didn't show Taiwan as part of China.

"China is not just 'at [Europe's] gates'—it is now already well within them." These words come from a study by the Berlin think-tanks MERICS and GPPI on the "authoritarian advance" of China's influence.[191] Uneasiness about China's behavior is breaking out in a number of places, both inside and outside Europe. In Vancouver—the destination of many well-off Chinese emigrants—people were more than a little surprised when the mayor put on a flag ceremony for the Chinese national holiday. The Chinese flag flew above city hall, while a band played the national anthem of the People's Republic.

At the end of 2017, the first congressional hearing

on China's "long arm" took place in Washington. Attempts by the Chinese government "to guide, buy, or coerce political influence and to control discussion of 'sensitive' topics"[192] ubiquitous in the West, announced the Congressional Executive's Commission on China. The heads of the CIA and FBI, too, have warned that the Chinese government has for some time been trying to increase its influence "on the whole of society," and "to a much greater extent than the Russians."

But nowhere else has the debate about China's influence started so early and so fiercely as in Australia. China is Australia's most important trade partner. Over a million of the country's residents have Chinese roots (out of a total population of 24 million), and half of them were born in mainland China. A recent investigation uncovered the pressure that Chinese student organizations—steered by Beijing—have been putting on university lecturers who criticize the motherland. The lecturers had broken taboos on topics like Tibet, Taiwan, or the Tiananmen Square massacre, and the student associations now demanded a change to the course and the teaching materials.[193] Australia's universities should give "consideration" to "the feelings of Chinese classmates," they argued—and pointed to the substantial amount of money that China's fee-paying students bring in every year.

Something else that emerged in this report was that Beijing had systematically forced the Australian Chinese community's media to toe the Party line, where until a few years previously it had been vibrant and diverse. Businesspeople loyal to China had also bought influence over members of parliament in Canberra, who then parroted Chinese propaganda word for word on sensitive topics, such as China's land-grab in the South China Sea.

There was an almighty uproar, which grew louder when the Australian author and academic Clive Hamilton revealed that a major publisher, Allen & Unwin, had canceled the publication of a new book he had written criticizing China. The manuscript was ready, but the publisher pulled out, fearing complaints from the Chinese figures named in its pages. No threats were ever made, says Hamilton. The fear of China alone—the "shadow it now casts over Australia"—was enough.[194] In the eyes of many Australians, the publisher's move proved exactly what the author had set out to argue—namely "that a powerful, authoritarian foreign state can suppress criticism of it abroad, and so smooth the path for its ongoing campaign to shift this country into its orbit." The book finally came out with another publisher, but the concerns it aired remain as pressing as ever.[195]

Foreign authors hoping to be published in China have long felt the impact of Chinese censorship: without removal or amendment of paragraphs or chapters that the Chinese state deems sensitive, publication is out of the question. A 2015 study by the PEN American Center shows how some authors comply, how others refuse—and how many are completely unaware they're being censored in the first place.[196] The study argues that by agreeing to censor their own work on topics like Tiananmen in 1989 or Tibet, foreign authors "further contribute to this historical amnesia" that the Party is enforcing on its people. Also, foreign publishers have long outsourced the printing of many of their books destined for their home market to China simply because printing there is cheaper. Recently numerous publishers discovered that suddenly they also had to agree to censorship in exchange for press time.[197] For example, books written by Australian authors and meant to be sold in Australia were no longer printed if they mentioned the names of dissidents like Ai Weiwei, if they treated certain religious issues, or if they contained maps that deviated from the official Chinese line that Taiwan is a part of the People's Republic and that most of the South China Sea is Chinese territory.

In an address to the national parliament, a representative of ASIO, the Australian intelligence service, called the "extent of the threat" from China's activities "extreme." There have also been warning voices in neighboring New Zealand. There, the public debate began when its secret service revealed that a Chinese-born MP, Yang Jian, had once taught future Chinese spies at an army school in Luoyang. The New Zealand sinologist Anne-Marie Brady is the author of *Magic Weapons*, one of the most detailed studies to date of China's attempts to exert overseas influence.[198] She believes that "China's covert, corrupting and coercive political influence activities in New Zealand are now at a critical level."[199]

After her ground-breaking report was published, Brady experienced what she described as a campaign of intimidation. Her house and office in Christchurch were broken into several times, the intruders clearly looking for her computer, mobile phone, and USB stick. She received anonymous phone calls and threatening letters ("You are the next"). Someone broke into her garage and her car was found to have been tampered with.[200]

China's efforts to influence the rest of the world aren't all in the form of covert operations. But "soft power," which relies on the natural charisma radiated by a nation and its culture, somehow isn't working for China. "Why is China so . . . uncool?" asked *Foreign Policy* magazine in March 2017.[201] Well, when was the last time the world saw a cool dictatorship? There's only so many points a nation can score with the invention of gunpowder and paper.

A survey carried out by the Pew Research Center found that in many African and Latin American countries the majority of people have a positive image of China. In the USA and Europe, those numbers are considerably smaller; the Germans in particular are notoriously skeptical. Ultimately, though, this doesn't matter, a US-educated economist at Beijing's elite Tsinghua university tells me: "We simply buy our influence. We've got enough money. Look at all the people who are already saying what we want to hear. And in the future, we'll be even richer."

In reality, Beijing employs a mixture of enticements and threats, recruitment and infiltration. As early as 2015, the American sinologist David Shambaugh estimated that China was spending around $10 billion a year on such image- and influence-boosting initiatives—far more than the USA, Britain, France, Germany, and Japan put together.[202] It's a collective effort, by no means confined to the diplomats: universities, think-tanks, friendship societies, and secret services are all in on the act, as well as Chinese companies that trade globally (including private companies). And of course, the Party and its multifarious departments are never far away.

One of the most interesting of these organizations resides in the heart of Beijing. West of the Gate of Heavenly Peace, where a portrait of Mao still greets you, lies the Zhongnanhai (Middle and Southern Seas) complex, which is forbidden to outsiders. It is the part of the old imperial garden that Mao's revolutionaries selected as their seat of government, immediately after the Revolution. This is China's Kremlin—and just next door, at 135 Fuyou Street, are the offices of the United Front: the organ through which the Party plans to bring Chinese people living abroad into line.

The United Front is an invention of Lenin's, and has been a staple of communist regimes all over the world. Traditionally its task has been to subsume new social groups, to neutralize troublemakers, to reel in persons of influence, and turn them into mouthpieces for the Party line. In China, the United Front looks after the eight political parties permitted for form's sake (as in the GDR, for instance) to exist outside the Communist Party. It also brings ethnic minorities, religious groups, and celebrities into the system. Billionaire entrepreneurs, pop stars and starlets, along with the odd Tibetan, Mongolian, or Uighur—always clad in splendid traditional dress—are invited to "political consultative conferences," whose primary purpose is to applaud the leadership. Now and then the guests may also make suggestions, but they have no real power or influence. The bosses of the new internet companies and start-ups are also regularly summoned to United Front seminars, where they are instructed in patriotism and "revolutionary traditions."

During the decades of reform and opening-up, China's United Front steadily lost influence and became a siding in which to park comrades who had earned an easy job. Now Xi Jinping has awoken the United Front from its slumber. He is calling it, as Mao once did, a "magic weapon" in the CCP's fight for influence. Today, according to Anne-Marie Brady, the organization "has taken on a level of significance not seen in China since the years before 1949."[203] One passage from a United Front handbook that fell into the hands of the *Financial Times* reads: "Enemy forces abroad do not want to see China rise and many of them see our country as a potential threat and rival, so they use a thousand ploys and a hundred strategies to frustrate and repress us."[204] Fortunately, though, the United Front is "a big magic weapon which can rid us of 1,000 problems in order to seize victory." As a strategy for operations on foreign soil, the China United Front Course Book recommends the usual mixture of charm and intimidation. It encourages functionaries to appear friendly, pleasant, and winning, in order to "unite all forces that can be united."

At the same time, they must stop at nothing to build an "iron Great Wall" against all "enemy forces."

For decades, the United Front preferred to operate in secret, but in 2018 the Party decided that there was no longer any need for it to hide behind other government agencies. The State Administration for Religious Affairs and the Overseas Chinese Affairs Office have been swallowed up, and the United Front now operates in its own name. Xi Jinping has declared overseas work to be a priority. At the start of 2018, he created a new steering committee, forging a direct link between the United Front and the Party leadership, and thereby signaling its central role in his future plans.

Among the tasks of the United Front is turning the residents of Hong Kong and Taiwan into fervent patriots. Other highly sensitive matters also fall within its remit: it is leading discussions with the Vatican about returning China's Catholics (more than ten million people) to the bosom of the holy Roman Catholic Church. This is something Rome has long desperately wished for, but has always been stymied by the Communist Party's insistence that the highest authority on earth for every Chinese citizen, including Chinese Catholics, is the Party and not the Pope.

This isn't the atheist Party's only power struggle with religious authorities. For some time, the United Front has been wrangling with the Tibetan government in exile over the reincarnation of the current Dalai Lama, who lives in India. The religious leader of all Tibetans has said a number of times in recent years that he has absolutely no desire to be reborn inside China. Beijing has reacted with obvious panic. A reincarnation of the Dalai Lama outside its control would be a nightmare for the CCP. The Dalai Lama can say and do what he likes, says Norbu Dunzhub, a United Front official from the "Autonomous Region of Tibet," but he cannot deny that "only the central government in Beijing" has the right to confirm his rebirth as genuine.[205] The current regulations for the system of reincarnation, the official reminded the Xinhua News Agency, had been "clearly set out in the relevant document of the state administration for religious affairs from the year 2007."

United Front workers, frequently in the guise of diplomats, also recruit Chinese businesspeople abroad, often through specially founded friendship and cultural associations. Western academics who show willingness are courted during generously financed conferences and trips to China; critical voices, on the other hand, may find themselves threatened with a ban on visiting the country. For many sinologists, that would mean the end of their careers.

Meanwhile some universities—including the prestigious Göttingen University in Germany—have Chinese professorships that are directly funded by Beijing. Indirect finance is also popular. When the University of Cambridge set up a professorship for "Chinese Development," it swore that the £3.7 million donation came from the private Chong Hua Foundation, which had no link to the Chinese government—until journalists revealed that the foundation, registered in the Bahamas, was controlled by Wen Ruchun, daughter of the former Chinese premier Wen Jiabao.

In 2019, the London School of Economics waived a plan to set up a China program with a multimillion-dollar donation from Shanghai businessman Eric Li only after an outcry from a large group of academics. Eric Li has been a poster boy of the Party-friendly press for years, preaching the superiority of the authoritarian model for China and defending the 1989 Tiananmen massacre. The same year, the United Kingdom's Parliament's Foreign Affairs Select Committee published a report that spoke of "alarming evidence" of Chinese influence at UK universities. Also in 2019, the German government informed that nation's parliament that it had "knowledge that Chinese authorities are trying to influence the actions of both Chinese students and researchers in Germany, and German researchers doing research in China."

The Communist Party carrying its ideological battles to the universities of the West is also increasingly a problem for universities that have opened a campus or research branch in China—expansions motivated by the fact that there is a lot of money to be made in China, but also because Chinese researchers have become increasingly intriguing as research and commercial partners. Of course China itself has also become one of the most important subjects of study today. At the beginning of 2020, there were fourteen American universities that had teamed up with local Chinese institutions (this kind of partnership is required by Chinese law if you want to set up a campus in China). New York University maintains a campus in Shanghai; Yale University operates a Center for Research in Cultural Sociology at Fudan University; Duke University has been a partner of the University of Wuhan since 2013; the University of California, Berkeley, works with Beijing Tsinghua University and the Shenzhen City Government; and Johns Hopkins University has set up a School of International Studies with the local university in Nanjing.

In December 2019, Shanghai Fudan University, together with Nanjing University and Shaanxi Normal University, caused a stir when it became known that all three had changed their charters under the supervision of the Ministry of Education in Beijing. Fudan University had removed "free thinking" and "academic independence" from its charter and replaced it with "patriotic devotion." Where the original charter had stated that the administration of university and teaching was in the hands of "teaching staff and students" pursuing their academic studies "independently and autonomously," the new version invokes "leadership by the Communist Party in the spirit of Marxism and socialism." The changes so outraged many Chinese students that a rare spontaneous act of protest took place on campus a few days after the changes became known: several dozen students came together for a flash mob and sang the school anthem, which had been composed in 1925 and whose words celebrate "free thinking" and "academic independence."

Of course, those who are involved in such Chinese institutions have had to make compromises in the past. "But these kinds of compromises were far easier to accept a decade ago, when a kinder, gentler version of the party ruled," writes the American journalist Bethany Allen-Ebrahimian, who has researched Chinese influence in the United States in recent years.[206] In view of the transformation of China under Xi Jinping, for Allen-Ebrahimian the pressing question now is: "At what point does engagement become complicity?"

Some have already decided to give up their China partnerships. For example, Cornell University abandoned an exchange and research program with Beijing Renmin University in 2018 after the university participated in the persecution and punishment of students who campaigned for labor rights.

When it comes to influencing academics and researchers, universities are not the only target of Chinese efforts. No matter whether in Washington or Brussels, new think-tanks tied to the CCP are springing up, so that Chinese academics or well-disposed foreigners can whisper the Party's message into the ears of the powerful. Generally, though, "the CCP's preferred method is to create dependencies and induce self-censorship and self-limiting policies," observes a study by the Royal United Services Institute, in a first attempt to present an overview of Chinese efforts to influence the UK. "The UK's departure from the EU may increase the CCP's desire to interfere," the study predicts, "as it seeks to implement further a 'divide and rule' strategy, aimed at imposing its global vision and promoting its interests."[207]

The Party strives to bring students living abroad into line through Chinese student associations. In Australia and the USA, increasing numbers of students are reporting that they are being spied on and subjected to "mind control" and intimidation by these associations, which are also active in the UK. Robert Barnett, one of the most prominent experts on Tibet, and director of the Modern Tibetan Studies program at Columbia University, reports that several Chinese students who had applied for research-assistant positions were, as he later discovered, sent by the Chinese consulate. And Wang Dan, the former student leader of the 1989 democracy movement, now says that "The Chinese Communist Party is extending its surveillance of critics abroad."[208] The Party is using a "campaign of fear and intimidation" to silence Chinese students in other countries. "Unpatriotic activities" like Wang Dan's seminars and salons at American universities are being recorded by "agents or sympathizers of the Chinese government," he claims, and Chinese students who attend are reported to the consulate. Those who overstep the mark run the risk that threats will be made to their families at home. Meanwhile, students at several US universities have set up Communist Party cells for the purposes of "ideological guidance," and on returning to their alma mater in China, some have been asked to report on their classmates' "anti-party thought."[209]

More than half a million young Chinese people are now studying abroad. Beijing is trying to organize as many overseas students and academics as it can, under the umbrella of the Chinese Students and Scholars Association (CSSA). In at least one case (at Georgetown University in Washington), it has been proved that the CSSA is in part financed directly by the Chinese embassy.

Sometimes student associations cooperate with their embassies to organize cheering, flag-waving crowds, as for Xi Jinping's state visit to Washington in September 2015. *Foreign Policy* reported that participating students were each paid $20 afterward.[210] The associations like to organize positive celebrations of "Chinese culture"—festivals such as the Chinese New Year—but if necessary they will also mount political protests. For example, when students from Hong Kong in Australia, New Zealand, Canada, or Germany demonstrate for the freedom of their hometown. Such protesters have been insulted and threatened by organized mainland Chinese, as has been the case in Hamburg, Sydney, and Toronto. Or, when the University of California in San Diego invited the Dalai Lama as a guest speaker. CSSA representatives insisted on a meeting with the university's rector, and agitated against the visit on social media platforms—doing their best to mimic and co-opt the ideas and terminology of American college activism. They created the hashtag #ChineseStudentsMatter, and claimed that inviting the "oppressive" Dalai Lama was an affront not only to the "feelings of the Chinese community" but to "diversity" and "inclusion." In this case, the university stood firm and the Dalai Lama spoke—unlike a few years previously, when North Carolina State University canceled his invitation, seemingly following an objection from the Confucius Institute embedded in the university.

Turning Chinese citizens living abroad into foot soldiers for the Party's cause is not the United Front's only purpose. Increasingly, it is attempting to recruit people of Chinese descent, even if they are no longer citizens of the People's Republic. It tries to appeal to a pan-Chinese patriotism, a pride in China's tradition and mission—the noble executor of which, of course, is the CCP. The 2018 amendment to the Chinese constitution included "patriots who endeavor to revitalize the Chinese race" in the list of groups forming a united front "under the leadership of the Communist Party of China." The *People's Daily* promptly published a lead article in which it invited "sons and daughters of China" all over the world to share in the prosperity and glory of the People's Republic's project.[211]

This appeal to racial solidarity, increasingly common in propaganda, is provoking dissent in overseas communities of Chinese descent. "When a state weaponizes patriotism and claims ownership of a people, it messes up [sic] with one's sense of identity, when identity is already a messy concept for ppl who crossed borders and linguistic barriers," read a widely-shared tweet by Yangyang Cheng, a physicist who lives and teaches in the USA. She told the story of how she herself was once called upon by the CSSA to cheer for China's president during a state visit ("need everyone to outnumber Free Tibet etc. protesters"), and spoke of a "cunning and ruthless state" exploiting the "people's vulnerability."[212]

There is also growing unease at the feeling that there is nowhere in the world where the CCP's eyes cannot see, or where its arms cannot reach. Its range has extended far beyond the lecture theaters and seminar rooms of American, Australian, and European universities. The kidnapping of the Hong Kong publisher and Swedish citizen Gui Minhai, who vanished from his vacation home in Thailand and resurfaced in a Chinese prison, may be the most spectacular example of the new cross-border activities of the Chinese security apparatus. The Gui Minhai case is interesting also because it showcases the increasingly assertive, coercive, and threatening tone of Chinese diplomacy toward countries that don't fall in line, Sweden in this case. The Chinese ambassador to Sweden, Gui Congyou, has long tried to intimidate the press, civil society, and politicians in Sweden into silence over the fate of the disappeared Swedish publisher. After the Swedish PEN in November 2019 awarded its Tucholsky Prize to Gui Minhai, and Swedish minister of culture Amanda Lind announced she would attend the award ceremony, as has been tradition for ministers of culture for many years, the Chinese ambassador escalated his threats. "We treat our friends with fine wine," he said in maybe the most astonishing of a whole line of astonishing interviews with Swedish media, "but for our enemies we got shotguns."[213] Later he threatened "countermeasures" and economic consequences for Swedish businesses.

The Gui Minhai case is a long way from being the only such instance of China disregarding borders on behalf of its suppression of dissent. Civil rights activists and dissidents living in exile, along with Tibetans and Uighurs, are also being targeted by Chinese agents, and suffering intimidation and threats.

These threats mainly relate to family members who are still living in China. One example is the journalist Chang Ping, already mentioned above. A critic of the government, he was fired from his editorial job at the Guangzhou paper *Southern Weekend*. Today he lives in exile in Germany, where he writes for the news service *Deutsche Welle*. Following the open letter he wrote to Chinese dissidents in March 2016, calling on

Party chief Xi Jinping to step down, his brother and sister in Sichuan disappeared into police custody for a few days. No, they had not "abducted" the pair, the police said; they were just questioning them about a "forest fire." At the same time, Chang Ping received a message from his brother, begging him to take back the letter and not to publish any more articles criticizing the Party. Later, semi-official websites in Sichuan published an open letter from the brother, which said he was "very disappointed" that Chang Ping wouldn't stop "spreading lies" from Germany and "maliciously attacking the authorities."

Uighurs living in France—including French citizens—have spoken about being approached by Chinese police officers asking for personal documents. The officers generally made contact via phone or WeChat. They demanded their work and home addresses, current photos, scans of their French ID cards, and even marriage and degree certificates. "Hello, I am a police officer with the [redacted] police station," *Foreign Policy* quotes from one WeChat message. "Let's have a good talk, otherwise it will be a lot of trouble to have to pay a visit to your father and mother's house every day."[214]

Meanwhile, China is offering substantial sums of money to bolt Confucius Institutes onto Western universities—and many are finding it difficult to say no at a time when finances are tight. Shortly after taking office in 2013, Xi Jinping made a speech in Qufu, the wise man's birthplace, in which he sang the praises of Confucius, whom Mao had condemned in such harsh terms. Xi lauded the Party as the protector of the old values, but he reserved his most rapturous praise for the Confucius Institutes, which were springing up like mushrooms all over the world. The success of the Confucius Institutes, said Xi Jinping, showed how hungry the West was for Chinese wisdom: "Their theory that capitalism is the highest good is teetering, while socialism is experiencing a miracle." The West's self-confidence had taken a knock, he claimed. "The countries of the West are comparing themselves openly or secretly with China's political and economic path."

In 2004 the Communist Party decided to turn Confucius into their Goethe, their Cervantes—the figurehead for a national cultural institute with a global mission. Today, there are more than 500 Confucius Institutes in 142 countries (at the end of 2019 there were 88 in the United States) and more than 1,000 Confucius classes. The government claims that almost 13 million people took part in the Institutes' activities in 2017. Confucius is attractive again—a brand. And wherever you see him, the Party's eyes peer out from behind the mask. How could it be otherwise, when the institutes that carry his name are "an important element of China's soft-power efforts abroad," according to politburo member Li Changchun? The Confucius Institutes like to make out that they are the Chinese version of the British Council or the Goethe-Institut, but there are good reasons to believe that, at heart, they are something quite different. To quote Li Changchun again: "It makes perfect sense to penetrate via the Chinese language."

The Institutes' Beijing headquarters is the "Office of Chinese Language Council International," known as Hanban for short in Chinese. Hanban's website says that the organization is "affiliated with the Chinese Ministry of Education." But that is only a half-truth: in reality, Hanban is controlled by a directorate that oversees over a dozen offices, including the propaganda ministry. Until her retirement in 2018, the directorate was led for many years by the Vice Premier Liu Yandong—a member of the CCP politburo.

The model on which the Institutes are set up is fundamentally different from that of their European counterparts. British Council offices and German Goethe-Instituts operate as independent organizations with their own premises. Confucius Institutes, on the other hand, are generally founded in cooperation with local universities. The British or German or French university provides premises and infrastructure; Beijing provides money through Hanban; a Chinese partner university provides teaching staff; and each side appoints its own director. A few years ago a debate kicked off: were our universities allowing themselves to be bought by an authoritarian regime? Were the Institutes, as one German sinologist has argued, really "Chinese submarines"? One thing is certain: these universities have invited an arm of the Chinese state to come in and make itself at home. Nothing comparable has ever existed.

Given that Hanban's core business is dialogue with the rest of the world, it is quite difficult to get hold of anyone there. My first requests for an interview at the start of 2014 went unanswered for nearly two months, and at some point they stopped picking up the phone for weeks at a time. Finally, Zhang Jing from the Institute for the Dissemination of Chinese Language and Culture said she was prepared to speak to me. This institute gives strategic advice to Hanban on the alignment and expansion of the Confucius Institutes. Their most important aim, said Zhang Jing, who herself studied in the USA, is to "feed the ideas of China into Western discourse." After all, she said, sinology had been founded by Western missionaries, and was dominated by the West to this day. "We in China were always the object of these studies. That carries with it the danger that you are either presented in an overly positive light, or in a distorted way." Beijing is currently less concerned about the over-positivity. In any case, Hanban is pleased to be able to give financial help to sinologists the world over: "We know that things are getting more and more difficult for the humanities everywhere. As a growing nation, China now has the means to support these studies."

Reports of censorship and self-censorship are coming in, particularly from the USA, where there are more than a hundred Confucius Institutes. In Los Angeles a few years ago, when Hanban announced that it was opening twenty institutes in collaboration with the US College Board, the Board's president, David Coleman, fell into poetic raptures: "Hanban is just like the sun. It lights the path to develop Chinese teaching in the US. The College Board is the moon. I am so honored to reflect the light that we've gotten from Hanban."[215]

Meanwhile a report by the National Association of Scholars, a conservative lobby group, has recommended shutting down all the Confucius Institutes in the USA. A few universities have already taken this step, both in the USA and elsewhere. In Chicago, Philadelphia, San Francisco, San Diego, Stockholm, Leiden, and Lyon, institutes have been closed after heated public debates. In the United States, twenty-nine Confucius Institutes have now been closed again, some most recently because of new rules stipulating that universities with Confucius Institutes will no longer receive funds from the Pentagon.[216] Germans like to stress that Confucius Institutes are set up differently in their universities, with more of an arm's-length relationship (unlike in the USA, where they are integrated into teaching programs, and can even award course credits). "There's so much more potential for conflict in the USA," says Philip Clart from the University of Leipzig. The institutes in Germany are almost all registered voluntary associations (e.V.s), which concentrate on providing language and cultural tuition to the general public, and are separate from the official day-to-day business of universities.

That may be so, but it is clear that Hanban wants more than just to provide language and tai chi courses for people. Beijing has declared quite openly that its aim is a "new sinology." "Our institutes started worldwide with language courses, calligraphy, and dumpling-making," says Zhang Jing. "But that's too limited for us. We want the Confucius Institutes to bring our vision into the academic world, into sinology. Quite simply, people have a false image of our country. We hope that sinology will start to engage more with Chinese politics and Chinese economics, and we hope it will explain to people that we can't just follow the Western path."

If you believe Heiner Roetz, a sinologist from Bochum in western Germany, then Hanban is pushing at open doors. "The fact that institutional partnership is even possible has something to do with the underlying readiness of some people to protect the Chinese regime, however much there might be to criticize," says Roetz. "And then they say we should leave China be; it's a different culture. The real problem isn't the Confucius Institutes at all—the problem is this cultural relativism within sinology."

Sometimes Hanban doesn't make things easy for its friends. At the end of July 2014, when Europe's sinologists met at the European Association for Chinese Studies conference in the Portuguese towns of Braga and Coimbra, Beijing managed to cause a substantial scandal. Xu Lin, Hanban's director general and a vice-minister-level official, had traveled from China specially for the event, which Hanban was co-sponsoring. When she discovered that one of the other sponsors came from Taiwan—and that this sponsor's logo appeared on the conference material—she seems to have given the order for all the conference programs to be collected. And so various unknown Chinese people marched along the rows, saying things like "Excuse me, I've forgotten mine" as they snatched programs out of the hands of baffled delegates. The following day, the programs turned up again, with the pages that had offended Beijing torn out. "It was a huge shock," says Daniel Leese, a historian and sinologist from Freiburg, who attended the event. Afterward, the EACS president Roger Greatrex published an angry letter objecting to this act of "censorship," and calling the behavior of the Confucius Institute's director "totally unacceptable."[217]

The huge costs involved in running the institutes have drawn criticism within China as well. But Xi Jinping seems to have even bigger plans for them. In a meeting of the Central Committee for the Deepening of Reform chaired by Xi at the start of 2018, a paper was passed on the reform of the Confucius Institute. It can be found on the website of the *People's Daily*, and contains the extraordinary declaration that in future, the institutes should better serve "the diplomacy of a powerful China."[218] The Party, then, explicitly sees the institutes as tools for exerting its influence. In October 2019, the Belgian authorities canceled the visa of Professor Song Xinning, head of the Confucius Institute at the University of Brussels. The accusation: espionage. Song was being accused of having actively recruited students for Chinese intelligence work from student and business circles. Song denies the allegations. And in an interview on Swedish radio, China's ambassador to Stockholm said that of course teachers at the Confucius Institutes had "the duty" to inform their students and colleagues about "the facts" of the Tiananmen Square massacre. In light of the clear message coming from Beijing, the loyal defense of the institutes by staff at partner universities in the West is at best seriously naïve.

As we have seen, the Party also has its eye on the West's academic publishing houses. Doing business in China is risky, especially when the coin you are trading with is academic freedom. Cambridge University Press (CUP), which describes itself as the world's oldest publisher, hit the headlines in summer 2017 after performing the Beijing censors' job for them. In response to questions from journalists, CUP confirmed that, at the request of the Chinese government, it had removed more than 300 academic articles and book reviews from the website of the high-profile publication *China Quarterly.* All of a sudden, many of the best-known sinologists in Europe and the USA found their work blacklisted. In a statement, CUP said it was cooperating with the censors in Beijing in order "to ensure that other academic and educational materials remain available to researchers and educators in this market."[219] A lucrative market, one might add: like other publishing houses, CUP earns good money from selling its publications to China.

At first, the Chinese Studies community was stunned into silence. Then came an outcry. Condemnation poured in from all over the world. Academics, journalists, and civil rights activists called the publisher's actions "disgraceful"—a "capitulation." "Pragmatic is one word," one professor tweeted; "pathetic more apt."[220] The general feeling was that CUP had sold its soul. "Scholarship does not exist to give comfort to the powerful," wrote MIT's Greg Distelhorst and Cornell's Jessica Chen Weiss in an open letter. "The censored history of China will literally bear the seal of Cambridge University."[221] James A. Millward from Georgetown University argued: "This is not only disrespectful of CUP's authors; it demonstrates a repugnant disdain for Chinese readers."[222]

In the end, CUP reversed its decision and released all the blocked material. It made a statement praising academic freedom as the highest good, and stating in its defense that the censorship had been a "temporary decision." Even after the furor had died down, though, the suspicion remained that this would happen again. Along with the premonition that each time it did, the outrage would become a little more muted, as censorship slowly became the new normality.

A few months later, the whole rigmarole was played out again in Berlin, at Springer Nature, one of the world's largest academic publishers.[223] On websites accessible from China, more than 1,000 articles featuring keywords like Taiwan, Tibet, Xinjiang, Cultural Revolution, or Tiananmen Square had been blocked. Springer Nature reacted rather differently from its British counterpart, shrugging off criticism from its own authors and maintaining its blockade on the articles that had offended Beijing. A few days later, Chinese state media celebrated the agreement of a "strategic partnership" between Springer Nature and the Chinese internet giant Tencent. Berlin responded to inquiries with a dry press release: "The blocking of content in China on the one hand and the cooperation with the Chinese internet company Tencent on the other are entirely unrelated." As of May 2018, according to its website, Springer Nature publishes more than 110 academic journals in cooperation with Chinese partners.

China is more than happy to shell out cash if it helps buy the cooperation of Western elites. It is constantly creating new foundations so that "friends of the Chinese people" can be given honorary titles and well-remunerated positions. "Elite capture" is the Royal United Services Institute's term for the mechanism by which these highly-paid consultants "risk becoming more amenable to CCP aims."[224] For example, Bob Carr, Australia's former foreign minister, headed the Australia China Relations Institute (which aims to spread "a positive and optimistic view" of Australia-China relations). Germany's former Vice Chancellor Philipp Rösler caused a stir when he turned up at the helm of a New York foundation part-owned by the mysterious company HNA, said to have close ties to the Beijing's "red aristocracy." After increasing public criticism, Bob Carr left the institute and Philipp Rösler left the HNA Foundation in 2019. David Cameron, who as prime minister proclaimed the "golden age" of British-Chinese relations, and still runs a British-Chinese investment fund, which intends to put a billion dollars into "supporting the 'Belt and Road' initiative" aka the "New Silk Road."[225] Clearly there is no shortage of willing cheerleaders for Beijing.

And in France, a former prime minister is working to spread the most positive image possible of the People's Republic of China and its Communist Party: Jean-Pierre Raffarin sits, for example, on the advisory board of the China-founded Boao Forum, Beijing's counterpart and rival to the World Economic Forum in Davos. Among other things, with the "Prospective et Innovation" foundation he manages, Raffarin organizes a large number of events sponsored by China built around themes such as the "New Silk Road." On the French TV broadcaster CGTN—actually the Chinese Communist Party's international propaganda channel—Raffarin has his own show, "Wide Angle on China." An episode entitled "Chinese Leadership: Full of Wisdom" begins with him in the opening credits leafing thoughtfully through Xi Jinping's book *Governing China*, and in the first part of the show Raffarin declares his admiration of the realm of Xi Jinping, "a powerful leader in a large country, where of course authority is required by one to rule four billion people." It is a country full of future and promise, led by a party with "a big vision" that "defines the rules of the future," a party that, in Raffarin's adulation, is dedicated to "protecting our planet," "international cooperation," and "defending multilateralism" for our common good. In 2019, Raffarin personally received the Friendship Medal from the People's Republic of Xi Jinping, a true friend of China. The Parisian newspaper *Le Monde* suspected that in Raffarin's case, his "motive was less the money than the recognition": Like other "friends of China," writes *Le Monde*, Raffarin was "a perfect mediator" for the party's united front strategy, which concentrates on "lobbying at all costs and influence."[226]

At the same time, the Party has mounted a global

media offensive. Its foreign broadcasters such as CGTN (television) and CRI (radio) have been allocated huge long-term budgets. In spring 2018, they were amalgamated with the national broadcasters CCTV and CNR to form a new entity: "Voice of China." The name bears witness to the continuing influence of the USA: first, the American Dream became the China Dream; now Voice of America has its counterpart in Voice of China. The new organization is one of the world's largest propaganda machines, and it is directly controlled by the CCP. According to Xinhua (which is itself increasing its overseas presence), its mission is to "propagate the theories, aims, principles and policies of the Party," as well as, most important, to "tell good stories about China." As of December 2018, the European headquarters of CGTN in London had advertised more than 350 new jobs for experienced TV journalists. According to media reports, the salaries were "a long way above" the London average.

CGTN claims to reach an audience of 1.2 billion in English, Russian, Arabic, French, and Chinese, which would make it the biggest broadcaster on earth. In video released in 2017, it argued that many outside China had been "brainwashed" by the "Western values of journalism"—a process that must be resisted. This from the country that currently occupies position 176 out of 180 on the Reporters Without Borders index of press freedom.

Year after year, hordes of African journalists are flown into China for "training." China provides many African editors with news content, technical equipment, and financial help in digitizing radio and TV channels. It is also laying fiber optic networks outside its own borders, opening data centers, and launching TV satellites into space as part of its "digital Silk Road." Helmut Thoma, former head of RTL (Germany's largest private TV station), coordinates the distribution of the China feel-good show *Nihao Germany*, co-produced by China's state news agency Xinhua, on German regional television. In the USA, the propaganda department has bought airtime on the Discovery Channel for the three-part celebratory documentary "Time of Xi," which according to the Party press "focuses on what China has to offer the world." The Beijing-based *China Daily* has also paid huge sums to include the propaganda supplement *China Watch* in high-profile Western newspapers like the *Washington Post*, *Le Figaro*, and (once) in the *Süddeutsche Zeitung*. The UK's *Daily Telegraph* reportedly earned £750,000 a year from it, until it stopped publication amid growing scrutiny of how Beijing was using its propaganda to shape the narrative of the coronavirus outbreak.

The *New York Times* defended the publication of the Chinese propaganda supplement in-house by pointing out that it had covered China "thoroughly and aggressively," and "at no time has advertising influenced our coverage."[227] In the case of the *New York Times*, that's true, and yet this argument overlooks an important point. Possible outside influence on reporting is only one thing—at least as important is the fact that *China Daily* and others use these ads to buy the reputation and credibility of Western media for the presentation of their propaganda.

On German television, *Nihao Deutschland*, a feelgood program co-produced by Xinhua, has recently aired its 500th episode. "Renting the boat to cross the ocean" is what they call such a strategy in China. Better still, says propaganda chief Liu Qibao, is buying the boat outright. Like the Beijing-loyal entrepreneur who has bought the Hong Kong *South China Morning Post*, or the others who are quietly purchasing shares in a radio network covering Washington and the surrounding area, as a Reuters investigation discovered.[228]

In central and eastern Europe, too, Chinese companies are bidding on several media consortia. The politburo member Liu Qibao oversees many of these activities. He is the head of the powerful Party organization that for many years was known as the "Propaganda Department of the Central Committee of the CCP"—and has now been re-christened the "Publicity Department" for the benefit of foreign ears. (In Chinese it still goes by its old name: *xuanchuanbu.*) Aware that changing the label wouldn't do the trick on its own, Beijing has also drafted in professional help from the West. PR firms like Hill+Knowlton, Ketchum, and Ogilvy are just a few of the big names who have helped the Chinese government spruce up its messaging.

For a while, the most visible such investments were in Hollywood, which has seen not only a wave of co-productions with China, but also a buying frenzy by Chinese firms, foremost among them Dalian Wanda Group. In 2012 the company bought AMC, the largest chain of cinema operators in the USA—then four years later, it bought Legendary Entertainment, the Hollywood studio that has produced films including the *Batman* and *Jurassic Park* franchises. Wanda's founder Wang Jianlin became China's richest man through real estate—and, like many other business magnates, he is very well connected with the Party leadership. In an interview, Wang put on record that his company naturally sees itself as contributing to China's "cultural influence."

Of course, the CCP would also love its domestic film industry to conquer the West with its artistic charisma—to go up against Hollywood's soft power. In spring 2019 the head of the National Film Bureau, Wang Xiaohui, set a high target for his country's directors: by 2035, China should have become a "strong film power" on a par with the USA. Wang showed his awareness that the quality of contemporary Chinese productions was sorely lacking: "Our ability to tell stories lags a long way behind Hollywood and Bollywood's," he told the assembled great and good of China's film industry. Yet this is unlikely to change any time soon—not because of a lack of talented Chinese directors and actors, but because of bureaucrats like Wang Xiaohui himself and the system he represents. The new generation of Chinese film, he told the artists as a parting shot, should be filled with "patriotic stories," and China's filmmakers must not "challenge the political system."[229] It was only in March 2018 that the Party placed the film industry under the direct control of the Communist Party's propaganda department. Here, as so often, the Party's paradoxical message is: castrate yourselves and be fruitful!

"Not since the Cultural Revolution have artists in China been as wary of the state, and as aware of the necessity of appeasing it," wrote the trade journal *Variety* in a report about the sudden disappearance in 2018 of China's largest film star, Fan Bingbing, which had shocked fans and the entertainment world.[230] With roles in films including the *Iron Man* and *X-Men* series, Fan had also become known to a Western audience. Over the course of four months, Chinese security agencies investigated her at a secret location, having apparently chosen the actress to serve as a warning example against income tax evasion. Fan Bingbing's disappearance made it clear to all Chinese actors how much they, too, are at the mercy of the almighty Communist Party. Not only did Fan Bingbing pay her tax bill plus penalties after she finally reappeared, but she has since regularly posted comments on her social media channels that emphasize her patriotism and loyalty to the Party.

In the Western film world, Chinese money remains more effective than Chinese charisma. Despite some sobering recent experiences (deals falling through, films flopping), today Hollywood is targeting blockbusters at the Chinese market and playing by its rules. The sheer size of that market makes temptation irresistible: the number of Chinese cinema-goers has grown by double-digit percentages in almost every year of the past decade, and at the start of 2018, China's box-office turnover outstripped that of the USA for the first time.

Hollywood has begun to take Chinese censorship into consideration, and this has certain concrete ramifications. The National Film Bureau in Beijing bans imported films from showing any content that attacks China's national pride, threatens its social stability, unity, and sovereignty, or shows the personalities of its leaders in a bad light. On co-productions, the NFB's regulations expressly require "positive Chinese elements."

When the Chinese director Jia Zhangke returned from a promotional tour for *A Touch of Sin*, he told me about a conversation he'd had with film journalists in New York. They had asked him if he had any tips for Hollywood producers on the smartest way to come to an arrangement with China's censors. "I was speechless," said Jia. "It churned me up inside." The actor Richard Gere, who is known as a critic of China and a friend of Tibet, told the *Hollywood Reporter* that he'd been kicked off projects because the Chinese side didn't want to work with him. Chinese baddies are cut from scripts, and invading Chinese troops (as in the remake of *Red Dawn*) are hastily transformed into North Koreans in the digital cutting room. At an earlier stage in the process, writers are increasingly adding Chinese heroes and heroines to their screenplays.

In Ridley Scott's 2015 film *The Martian*, for example, the China National Space Administration gets to help the American NASA out of a desperate situation with a supply rocket, saving the astronaut stranded on Mars (played by Matt Damon) from starvation. *The Martian* brought in $94 million in China. In real-life Beijing, Xu Dazhe, head of CNSA, cited the film as proof that "our American counterparts have high hopes that they will get to work with us." Regrettably, however, there were "a few obstacles" in the way of this cooperation—an allusion to the decision of the US Congress in 2011 to ban NASA from collaborating with Beijing in any way whatsoever, for reasons of national security.

Unlike Russia, China goes well beyond feeding nuggets of disinformation into the public discourse—it is turning its attention to infrastructure and institutions. The "thought work" perfected in China itself is to be outsourced on every imaginable level. The goal is a new world order.

China is pursuing its aims on the grandest global scale. Its diplomats are increasingly active on the United Nations Human Rights Council in Geneva. There they are increasingly successful in getting through their own resolutions—giving some of their harmonious-sounding, home-grown propaganda a place on the world stage. The resolution advocated a "community of shared future" and "mutually beneficial cooperation" in the field of human rights, and its sponsors included such states as Syria, Egypt, Myanmar, Burundi, and

Eritrea.[231] The wording sounds harmless enough, if a little flowery, but human rights activists see the resolution as a first step in an attempt to rewrite international human rights norms. "The Communist Party of China (CCP) has always strived to pursue happiness for the people and development for [. . .] mankind," wrote China's ambassador to South Africa in an article for the English edition of the *People's Daily*. "Today, the 1.3 billion Chinese people are enjoying their life of peace, freedom and happiness, free of fear of war or conflict. China, with its fastest human rights progress and best practices of human rights protection, is setting up a new model for the world cause of human rights."[232]

In this model, human rights are to be redefined, and political and civil rights are to be replaced by "economic" and "social" rights; it should therefore suffice if a state dresses and feeds its people. Unique national "characteristics" are sufficient to justify violations of human rights. The sovereignty of states is the highest of all rights. Yes, says China, there should be one kind of democratization—namely the "democratization of international relations." Meaning: in the future, on a subject like human rights China's word should be worth as much as that of the countries actually supporting in spirit the United Nations Universal Declaration of Human Rights.

China's hope is that the ideologically charged terms

of its propaganda apparatus will redefine international discourse. The Party, for example, hopes to also establish its Chinese-style "community of a shared future" or "common destiny" in the online sphere. The global internet, it believes, urgently needs reinventing. It should be secure. Orderly. Obedient. This is the purpose of the World Internet Conference in the southern town of Wuzhen. Every year since 2014, it has gathered politicians, academics, and business executives from all over the world, with the aim of spreading its vision of an internet that "serves national security" and "respects national sovereignty" around the globe. Wuzhen is the perfect venue: an old, painstakingly restored southern Chinese water town, with picturesque canals and bridges. Ancient China has been reborn here, as a lucrative theme park aimed at the tourist industry, cleansed not only of its former inhabitants, but also of dirt, trash, and any kind of life. It is separated from the outside world by a long wall, several guarded gates, and a wide canal. Today, it is a pretty, empty shell waiting to be filled with commerce and propaganda—every bit as open and vibrant as China's internet itself.

The conference, hosted by China's cyberspace authority, the CAC, is avoided by Western politicians. They have no real interest in the "community of shared future in cyberspace" advocated by Wang

Huning, the Communist Party's chief ideologue, in his opening speech in December 2017. The deputy leaders of Mongolia, Pakistan, and several southeast Asian countries, on the other hand, were all listening carefully. For them, China's "effective internet security management" is a source of admiration—and potentially "of use to all our countries," as the delegate from Thailand's military junta put it in his address.[233]

Their politicians may be absent, but Wuzhen has in some years become a place of pilgrimage for the Western tech elite: they are invited here by the regime, which offers them a gateway to the legendary Chinese market. In 2017, Apple boss Tim Cook found words of praise for his hosts; there was, he said, a "vision that we share." That same week, the Washington-based NGO Freedom House once again crowned China the "world's worst abuser of internet freedom," ahead of Syria and Ethiopia.[234] No doubt what Cook had in mind was the part of Wang Huning's speech where he said: "Let us make the pie of the digital economy larger!" Other attendees included top executives from Cisco, IBM, Google, and Facebook: the enticing aroma of this pie lures plenty of Silicon Valley giants—even those whose products are blocked and banned on the Chinese market.

While the executives talked about networking, openness, and—water-town metaphors being extremely popular in Wuzhen—building bridges, another event was taking place next door. In the Jingxing conference room, the Ministry of Public Security co-hosted a discussion on "international cooperation in combating the misuse of the internet for criminal and terrorist purposes." Unfortunately, said Professor Zhang Jinping, foreign media often used "the language of the terrorists." China needed to make more of an effort to fight back: "We must oppose it everywhere in the world with our own language. Our own narrative must gain the upper hand."

Professor Mei Jianning from Shanghai went on to illustrate the problem with an example: he named three reasons why the World Congress of Uighurs—based in Munich and recognized by the German authorities as a peaceful organization of exiles—should be classed as a "terrorist organization." First: "They twist the facts and use a polluted language." Second: "They say negative things about the national 'One Belt, One Road' project." And third: "They collaborate with other anti-China organizations." Professor Mei recommended that the Chinese government put more pressure on Twitter and other Western social media companies to block the accounts of the World Congress of Uighurs and other similar "terrorist" organizations.

The foreign guest speakers that day included a

Turkish police officer from Ankara and a liaison officer from the Iranian embassy in Beijing, who identified the US government as the greatest terrorist threat on the web. "It will be a bitter battle," the chair summarized at the end. "We need to catch up quickly."

Remarkably, after the US began sanctioning Chinese internet and high-tech companies in 2019, it was a Chinese politician who complained at the recent World Internet Conference in Wuzhen that "the foundation for an open and shared-by-all internet is unstable."[235] Huang Kunming, a member of the Politburo, which comprises China's 25 most-senior officials, said at a technology forum that "Some countries restrain and suppress companies from other countries using cyber security as an excuse"—mind you, he wasn't talking about his own country, but about the United States.

The number of partners on board China's journey to a closed, monitored internet is growing. Russia has proven to be a particularly keen student. Several creators of the Chinese censorship system have traveled to Moscow to pass on their experience. In 2016 the former head of the Cyberspace Administration, Lu Wei, and the architect of the Great Firewall, Fang Binxing, took part in the 7th International Safe Internet Forum. The forum is hosted by the Safe Internet League set up by the oligarch Konstantin Malofeev, a man with close ties to the Kremlin, the GRU (Russia's military secret service), and the orthodox church. Not long afterward, President Vladimir Putin and Xi Jinping held a summit at which they agreed to cooperate more on the internet and cybersecurity. For the Russians, acquiring cutting-edge new technology is the main prize: the Chinese firm Huawei, for instance, is to help the Kremlin create data storage systems.

Human Rights Watch has revealed that the Chinese telecoms company ZTE set up a telecommunications system for the government of Ethiopia, helping it to monitor activists, journalists, and members of the opposition.[236] Manufacturers of facial-recognition cameras and other AI technologies are increasingly exporting their wares. At the start of 2018 the company Yitu announced that Africa was the next big market, and as early as 2017 Huawei was claiming that its Smart City concept was active in more than 200 cities in 40 countries. Customers can order their Smart City with or without surveillance components—though the company's sales presentations suggest strongly that, in the interests of security, they should buy the whole package: "A safe city is the foundation of a smart city."[237]

There are also plenty of willing buyers in South America. According to a glowing feature in Xinhua, the China Electronics Corporation (CEC) has set up surveillance networks for the governments of Venezuela, Bolivia, and Ecuador.[238] CEC is a state-owned company that also makes products for the Chinese army. In Ecuador, CEC has installed a camera network on its integrated ECU911 security system, along with a system of smartphone surveillance. The crime rate immediately dropped by a quarter. The president of Ecuador at the time, Rafael Correa, thanked China for generously donating some of the surveillance equipment.

In 2019, China was once again named by Freedom House as the world's "worst abuser of internet freedom." According to the report, "global internet freedom declined for the eighth consecutive year"—and this decline could be traced directly back to the international spread of "the Chinese model of extensive censorship and automated surveillance systems."[239] Of 64 countries investigated by Freedom House, China had sold 38 its internet and telecommunications infrastructure, and it had exported surveillance technology with AI capabilities to 18. It had carried out training in the management of the internet and new information technology in 36 countries, and in Vietnam, Uganda, and Tanzania, among other places, this had led directly to new legislation based on the Chinese model.

In 2019, a study by the Open Technology Fund identified 102 countries to which China had exported information-control technologies. These included autocracies such as Egypt and Azerbaijan, as well as semi-authoritarian or even democratic states such as Brazil, Malaysia, Poland, and South Korea.[240]

In Zimbabwe, for example, Freedom House reports that the Chinese firm CloudWalk is creating a nationwide system of surveillance and facial recognition. As part of the deal, Zimbabwe has agreed to send the biometric data of millions of its citizens to servers in China, where CloudWalk can then train its algorithms to recognize faces with darker skin tones. "These trends present an existential threat to the future of the open internet and prospects for greater democracy around the globe," writes Freedom House. At the Party Congress in 2017, Xi Jinping presented his plan to turn China into a "cyber superpower." In the same speech, he said that the Chinese model was "a new option for other countries and nations that want to speed up their development while preserving their independence."

Any technology, including artificial intelligence systems and surveillance cameras, can be used for good or ill. But its increasing sophistication and efficacy make it a temptation for rulers of all stripes—and the Chinese export drive that is just beginning comes at a time when freedom and democracy are under threat, all around the world.

The Future

When All Roads Lead to Beijing

"If Europe was a person, I would have to
charge off and fight for her now. For my heroine,
who has given me 70 years of peace."

Klaus Maria Brandauer, actor

"Carry China's voice everywhere," Xi Jinping
urges his country's propaganda machine. The
West's power in the "global debate" must be challenged, and the window of opportunity is wide open:
"Democracy and pluralism are under assault," says the
Freedom in the World 2020 report by Freedom House;
the think-tank sees democracy in "the most serious
crisis in decades." A new report[241] by Cambridge University's Centre for the Future of Democracy comes
to a similarly disturbing conclusion: "Dissatisfaction

with democracy has risen over time, and is reaching an all-time global high, in particular in developed democracies." In the 1990s, the authors write, two-thirds of citizens in the democracies of Europe and North America had stated that they were satisfied with their country's political system. And just ten years ago, three out of four Americans still said they were satisfied with democracy. Not any longer apparently. "For the first time on record, polls show that a majority of Americans (55 percent) are dissatisfied with their system of government," the authors reported.[242] "This marks a profound shift in America's view of itself—and its place in the world."

Not the worst time for Chinese propaganda. For some time, the narrative of the munificent "New Silk Road" has been doing wonders for Xi Jinping. It would be more accurate to talk about a whole network of new Silk Roads, running in every direction, and all with one thing in common: they begin and end at the imperial palace in Beijing.

One branch links China by rail to the DP London super-port at Tilbury. Since January 2017, the "London-Yiwu" train has been plying the 7,500-mile route, laden with British and Chinese goods. It's just one limb of Xi's vast signature project, which the propaganda department renames every so often, so that it is also known as One Belt, One Road (OBOR) or Belt and Road Initiative (BRI). New roads, railways, ports, airports, and pipelines are to be built; new trade corridors to connect China with central Asia and Europe—but also to turn South America and Africa toward Beijing. So far, more than 80 countries have agreed to participate.

The plan is part economic program for Chinese industry, part global infrastructure project. First and foremost, though, it is a geo-strategic vision for a new world order determined by China. Long before the first building projects were begun, the propaganda was already working nicely. "China's 'One Belt, One Road' will be the new World Trade Organization—whether you like it or not," said the Siemens boss Joe Kaeser at the World Economic summit in Davos, once again standing out among his fellow German captains of industry for his skill in saying exactly what the Chinese authorities want to hear. Naturally he would like his company to get a slice of the pie.

Elsewhere in the EU, alarm bells are ringing. In spring 2018, 27 of the 28 EU ambassadors in Beijing produced a report arguing that the Belt and Road Initiative was China's plan to shape globalization according to its own design.[243] Not that China is denying it: the *People's Daily* optimistically prophesies a "globalization

2.0."[244] The diplomats' report warns, however, that the new system will be to Europe's disadvantage: the New Silk Road threatens EU interests and standards. Even when it comes to infrastructure projects, Beijing prioritizes political interests over economic considerations. The deals are negotiated exclusively through bilateral government summits—a tactic that enables China to exploit the "unequal distribution of power," write the ambassadors. So far, the tendering process has been as un-transparent as one might expect, and has shown a bias toward Chinese state firms. Companies that abide by European environmental, employment, and social standards are at a disadvantage. China is also demanding that states who want to participate acknowledge its "core interests," among which Beijing counts its territorial claim on the South China Sea.

It is telling that one of the 28 EU ambassadors refused to give his signature to the report. Not for the first or last time, Hungary was the country breaking ranks. For some time now, as China pursues its own interests, it has been deliberately sowing dissent among Europeans and the European Union. This is causing a growing sense of indignation in Brussels and in the national governments of the major EU states. In the last few years, China has succeeded in getting several joint declarations by the EU overturned or watered down—in particular via Greece, Hungary, and the Czech Republic, who have repeatedly acted as proxies for Beijing. In 2017, for example, at a meeting of the UN Human Rights Council in Geneva, Greece overturned an EU declaration that criticized China's rapidly worsening human rights record. It was the first time the EU had failed to get its members to agree on a joint declaration, and came just a year after the International Court of Justice in the Hague had ruled China's territorial claims in the South China Sea negligible.

Hard cash (or the tantalizing prospect of it) is helping to smooth China's entry into the heart of European affairs. The "16-plus-1" initiative, for example, has assembled a bloc of sixteen post-Communist central and east European countries into a kind of "China, please invest in us" club. The "16-plus-1" countries—in China the project is known more honestly as "1-plus-16"—meet once a year, when the participating states have an audience of a few minutes with China's head of government, hoping for billions in Belt-and-Road investment. The home of the 16-plus-1 general secretariat is the foreign ministry in Beijing—it is here that the deals are drawn up for the European members to then wave through. Some countries are still waiting in frustration for Chinese money, but others have begun to receive investment. In the meantime, the format has grown: Greece has joined and the alliance is now called "7-plus." Under Chinese management—and following more than four billion dollars of investment—the Greek port of Piraeus has become the fastest-growing container port in the world. In Hungary, China is intending to build a high-speed train, and the Czech Republic has been promised more than eight billion euros in investment by 2020. In April 2020, for the first time Xi Jinping himself wanted to lead the meeting instead of Prime Minister Li Keqiang, one of many signs that China is increasingly paying attention to the European continent in times of increasing rivalry with the United States.

The Czech president Miloš Zeman expressed his thanks in an interview with China's state broadcaster CCTV, in which he bizarrely promised to put an end to the Czech Republic's "subservience to the USA and the EU." Zeman referred to Xi Jinping as his "young friend," and the Belt and Road project as "the most remarkable initiative in modern human history." Even so, of the 3.5 billion euros promised to the Czech Republic by China in 2016, only 362 million were delivered. By comparison, the Czech Republic receives 5.7 billion euros a year from Brussels, making it the second-largest net beneficiary in the EU. Nevertheless, Zeman won the presidential election again in January

2018 after a campaign in which he repeatedly mouthed off about the EU, while painting China and Russia as more suitable partners.

Shortly after the vote in the Czech Republic, Zeman's Chinese confidant and influencer Ye Jianmin, head of the distinctly opaque energy company CEFC, met the fate of so many Chinese company bosses: the Chinese security apparatus spirited him away due to "legal violations." Prior to this, however, he had managed to buy a Czech airline, a brewery, a football club, and a media company—and, most notably, created a number of well-paid posts for current and former Czech politicians and advisors to Zeman. "As it turns out, CEFC's main investments in the Czech Republic weren't economic, they were about buying up the loyalty of Czech officials," writes Martin Hála, a Czech sinologist and director of a project called Sinopsis, which studies China's growing influence in the Czech Republic and neighboring countries. Instead of innovation, says Hála, China has given Europe "a new take on age-old crony capitalism."[245]

At the end of 2019, Charles University in Prague was shaken by a scandal: it emerged that four of the faculty members had been secretly paid off by the Chinese embassy in Prague, including Milos Balaban, head of the Center for Security Policy. "This shows us how vulnerable universities are to foreign influence," said a faculty spokesman.[246] Such revelations are now causing a backlash in some places. For his part, the new mayor of Prague, Zdenek Hrib, has made a sport of counteracting President Zeman's China-friendly policy: first he let the Tibetan flag fly at Prague City Hall, and then he not only cut the ties to Prague's twin city, Beijing, but to the horror of Chinese diplomats, made Taiwan's capital, Taipei, a new twin city.

But politicians like Hrib are the exception. Hungarian diplomats, as their EU colleagues report, are reading out statements on Beijing's human rights record that sound as if they were written by China's foreign ministry. "There is probably no other country that uses economic influence so directly to pressurize countries into political compliance," a European diplomat in Beijing told me. "With countries like Hungary and Greece, China is now practically sitting round the table in Brussels on these issues. There is no longer any central issue relating to China on which the EU can still agree."

When the European Union prepared a report on global disinformation campaigns in the wake of the coronavirus pandemic at the end of April 2020, China apparently succeeded in directly influencing the final version of the paper. "China has continued to run a global disinformation campaign to deflect blame for the outbreak of the pandemic and improve its international image," the initial report said, according to the *New York Times*, which had obtained a copy. Following China's intervention, this sentence simply vanished from the report, as did a section on slanderous Chinese attacks on French politicians, much to the frustration of the authors. "The Chinese are already threatening with reactions if the report comes out," Lutz Güllner, a European Union diplomat, wrote in an email quoted by the *Times*. And the EU officials responsible complied: Brussels self-censored under pressure from Beijing.

China's Communist Party has shown again and again that it believes it can flout the rules of the old world order if they aren't in its interests. It has simply ignored the decision from the Hague on the South China Sea, continuing to create new islands and enlarging existing ones into potential naval bases, as the militarization of its waters proceeds apace. The autonomy of Hong Kong, which China guaranteed in 1997 for fifty years, in treaties with the former colonial ruler Great Britain and with Hong Kong itself, is being suffocated after just twenty years because it is inconvenient.

The Party is feeding on the weaknesses of the West. Many Chinese people once viewed the West as a model of economic success and principles—both reputations have taken a hit in the last few years. The 2008 financial crisis was one watershed moment; Brexit, Trump's election, and the recent successes of right-wing populists in other Western countries have been others. The Arab Spring and Ukrainian revolutions, upon which the West looked so favorably and which finally failed so spectacularly, also provided rich fodder for Chinese propaganda: "Countries that copy Western democracy end up with hunger, poverty, chaos, and bloodshed," the Party press warned its readers.[247]

In spring 2020, at the height of the coronavirus crisis, the world witnessed an impressive spectacle: the Chinese Communist Party, master in the discipline of rewriting history, now tried this trick in front of the eyes of the whole world. The collective amnesia so often inflicted on the Chinese this time was to infect the rest of the world as well. With astonishing chutzpah, China's foreign propaganda harnessed the virus for itself in the competition between systems. In their story, all of a sudden, China was indeed the role model, the country that "can pull together the imagination and courage needed to handle the virus, while the US struggles" (in the words of the *People's Daily*). In other words, China was in fact the country with the superior political system and therefore had what it takes to become the new global leader. China owes the world no apology, Xinhua wrote in early March, quite the contrary: "The world owes China thanks." As a reminder, this is the same China that had concealed and covered up the outbreak for weeks, which is why before Wuhan's lockdown, five million people were able to leave the city untested and spread the virus to the rest of the country and the world. The same country whose reported infection figures are still mistrusted by many serious scientists today.

In the propaganda battle that followed, China for the first time also relied on Russian techniques of disinformation on a large scale: official Chinese Twitter accounts spread conspiracy theories about the origin of the virus. One of them said that the virus did not come from Wuhan but from a US bioweapons laboratory—that theory was shared on Twitter by one of the spokespersons for the Chinese Foreign Ministry, Zhao Lijian. This disinformation offensive was accompanied by a well-coordinated "mask diplomacy": all over Europe, Air China planes with tons of face masks and other medical equipment touched down, cheered on by, among others, many thousands of remote-controlled Twitter bots. The public only learned later that many of those masks were not being donated, but rather sold. Still, the authoritarian EU skeptics among Europe's leaders in particular took this as an opportunity to publicly praise China as a savior in need: Hungarian prime minister Viktor Orban personally welcomed the Chinese aid plane, and Serbian President Alexander Vučić kissed China's flag and said: "China is the only friend who can help us."

Of course, states such as Taiwan, South Korea, and even Germany showed that democracies can fight the virus at least as successfully and more humanely than the authoritarian state, if they rely on scientific expertise and political determination and can be sure of the trust of their people. But, as so often happens, the West also unwittingly fed into China's narratives. The Europeans, for example, have long forgotten how important narratives and symbols are in the political project; their response to Beijing's massive propaganda efforts so far has been terribly pale and feeble. More important, however, in the beginning of 2020 European states were far too complacent for too long. And then, when the pandemic hit with full force, shocked Europeans all retreated into the nation state, forgetting at first to show solidarity with badly hit neighbors like Italy. But China's propaganda media were even more pleased about the ever-erratic US president Donald Trump, whose ignorance and hostility toward science soon made the US the center of the pandemic: the Communist Party could not have wished for a better partner in its plan to make its own sins forgotten.

The West's betrayal of its own ideas has been playing into the hands of the autocrats for some time. This began long before Donald Trump. The US soldiers who tortured prisoners in Abu Ghraib; the prisoners in Guantánamo stripped of all rights; Edward Snowden's revelations: all this has been music to the ears not just of conspiracy theorists but also, more important, of the world's dictators. The USA is the self-declared beacon of freedom and democracy, a vocal supporter of transparent government, checks and balances, the free flow of information. Was it now using the same police-state tactics as the regimes it always condemned? Naturally, Washington claimed it was only trying to protect its people from terrorists—but that's what Beijing says, too. A stabbing in the Uighur province of Xinjiang? Terrorism! The Dalai Lama praying for Tibetans who have set themselves on fire? Terrorism in disguise!

On it goes, week after week. The USA wants all visa applicants to hand over their social media passwords? Bravo, cries the state press in Beijing, and thank God the world is increasingly "united" on the subject of internet management.[248] Donald Trump going on about fake news again? The *People's Daily* bows in gratitude—and it does so on Twitter, so that he can read it too: "@realDonaldTrump is right. #fakenews is the enemy. China has known this for years."

Cynics everywhere are rubbing their hands: isn't this what we've always said? Aren't they all the same: Washington, Beijing, London, Berlin, Moscow? Of course they aren't—but must we make it so easy for the lazy thinkers and the autocrat-appeasers? Is it really necessary to betray the cause of freedom and human rights ourselves, a little at a time?

Barack Obama and his government defended the surveillance practices of their secret services with the argument that they were legal. As if that very fact were not scandal enough: all the things that have become legal in our countries. Look at what US governments before Donald Trump's have considered lawful. The imprisonment of suspects without charge and without trial. The torture of prisoners. The killing of suspects in other countries using unmanned drones. What will the USA do, and what will we do, when the first Chinese drones start killing Tibetan "terrorists" in Nepal or northern India?

It may be one of the great ironies of human nature that—imperceptibly at first—we often become more and more like our enemies. But we don't have to capitulate. We must preserve our sense of alarm, not allow ourselves to be numbed by our daily horror at what is happening around us. We must try not to dodge the fist that punches us in the guts first thing in the morning as we read the news.

In the end, rather than just pointing the finger at China, we need to look at ourselves. Yes, China is trying to divide Europe, but you can only divide something that is already divisible. We need to wake up: pinch ourselves, open our eyes, and take a hard look in the mirror—paying special attention to the nasty bits that are disfiguring us beyond recognition. We need to be aware of the dangerous attractions of the high-tech future, tempting us to betray ourselves and our values.

China, with its AI and big data projects and its Social Credit System, can serve as a useful example. Developments in China are being replicated here—so far in a rudimentary way here and mostly in commercial contexts—whether in the realm of facial recognition, behavior prediction, or online reward-and-punishment ecosystems. China can be a mirror in which we see our dark selves reflected; and the single-minded ruthlessness with which the Party is putting these new technologies to use is giving many a CEO in Silicon Valley, London, or Berlin wet dreams. (For years the suggestion that we urgently need to emulate the Chinese in one area or another, or we will fall terribly behind, has been a standard complaint from a certain type of executive for many years.)

It's crucial, too, that we are clear-sighted about all the ways in which the CCP is trying to influence and change us. Russia currently dominates the conversation in the USA and Europe—but the greatest challenge to liberal democracy will come not from a stagnant Russia. It will from the economic powerhouse of China.

Of course, the West will keep talking to Beijing, and continue to do business with China, but it is our responsibility to keep our eyes open as to the true nature of this regime. Every country has a right to promote itself and to fight tooth and nail for its own interests. Why should China be shut out of the global exchange of ideas? Yet we should remember the differences. China's leadership exploits the opportunities afforded by an open society to extend its reach, but does not act in an open and transparent way itself. Where it exerts its influence, it is in the service of an authoritarian system, whose values and practices it aims to spread beyond its own borders. It attempts to infiltrate other countries' institutions and weaken them, where that serves its own interests. It often works covertly to undermine pluralism and freedom of opinion. And it does all these things with the conviction that it is fighting an ideological war with the West.

The challenges for the West and for liberal democracy are manifold: on the one hand, there is the destructive power of Donald Trump and the apocalyptic noise of right-wing populists within our own midst; and on the other hand, there are Russia and China. There is a perfect storm brewing. China is just one piece in this puzzle, though it is thus far the most underrated.

Beijing relies on the weaknesses and disputes afflicting our democracies, so preoccupied with their own navel-gazing that they have almost stopped seeing what is going on around them. The betrayal of democratic values by Daimler, Apple, and Springer Nature are just the beginning. Many more will follow, often so banal that they will pass unnoticed, but they will be no less poisonous for that.

There is a battle ahead, and first and foremost it must be fought at home, among ourselves. We Europeans—we exhausted, quarrelling, complacent, inward-looking, lame Europeans—need to see China for what it is. And on the other side of the Atlantic, the Americans need to wake up to the autocratic threat from within and join ranks again with their democratic allies around the globe. We need to act, to throw off our mantle of narcoleptic naivete, and reunite, to rediscover the strength and brilliance of the ideals so many generations fought for. And our ideals are strong—that's the good news. Despite all its swagger and bluster, the Chinese Communist Party is very much still afraid of the allure of Western democracies and Western values. Above all, it is this fear that lies behind all the efforts the Party is making to influence the West.

A competition of systems has returned. Will China overtake the West and surge ahead to lead the world? The answer will depend in part on China's strength, but even more it will depend on the West's weakness. As dictatorship is reinventing itself before our very eyes in Beijing, yet perhaps our most pressing task in the West—from Berlin to Athens, Prague to Paris and, yes, in Washington and in London, too—is the reinvention of the West, the reinvention of Europe, the reinvention of democracy.

In the end, it will not be decisive how strong China is; what will be decisive is how strong we are. Or rather: how weak we are, how much we split, how much we let ourselves fall into fatalism and resignation. We still have the better cards. We just can't let them take them out of our hands.

And no, we don't have to fear China, we just have to fear ourselves.

Acknowledgments

Thanks are due to China, for the endless adventures it has given me over the past three decades. And to its people: those who have become my most loyal friends, and those who shared their stories and insights with me without having met me before. They sometimes took great risks to do so, and many had to endure harassment as a result. For that, I would like to apologize.

I would also like to thank my family for their great patience during my work on this book. Thanks to Bernhard Bartsch, who was the first to look through the manuscript and provided valuable tips. And to my editor Martin Janik, who retained a constant overview and was always there to provide the necessary impetus at the right time.

They have all made this a better book. But its failings are my own.

Notes

1 James Mann, *The China Fantasy: Why Capitalism Will Not Bring Democracy to China*, London 2008.

2 For more on the Communist Party's appetite for experimentation and its ability to change, see Sebastian Heilmann, *Red Swan: How Unorthodox Policy-Making Facilitated China's Rise*, Hong Kong 2018.

3 Stein Ringen, a Norwegian scholar of sociology and political science based at Oxford University, calls the CCP's rule "the perfect dictatorship" and terms it a "controlocracy": "Although the controlocracy is sophisticated and does not depend on the omnipresence of terror, the threat of terror is omnipresent, and that threat is backed up by a physical use of violence that is sufficient for citizens to know that the threat is not an idle one." (Stein Ringen, *The Perfect Dictatorship: China in the 21st Century*, Hong Kong 2016, pp. 139–140.)

4 A representative example here is the exposé in the *New York Times* about the family of Wen Jiabao. Until 2013, Wen was China's prime minister, and state propaganda always painted him as the modest, folksy "Grandpa Wen." The *New York Times* revealed that by the end of his period in office, his family had amassed a fortune of at least $2.7 billion. See David Barboza, "Billions in Hidden Riches for Family of Chinese Leader," *New York Times*, October 25, 2012 (https://www.nytimes.com/2012/10/26 /business/global/family-of-wenjiabao-holds-a-hidden -fortune-in-china.html).

5 Li Laifang, "Enlightened Chinese democracy puts the West in the shade," Xinhua, October 17, 2017 (http:// www.xinhuanet.com/english/2017–10/17/c_136685546 .htm).

6 Extract from interview with Hannah Arendt by the French writer Roger Errera, "Hannah Arendt: From an Interview," *The New York Review of Books*, October 26, 1978 (http://www.nybooks.com/articles/1978/10/26/hannah -arendt-from-an-interview/).

7 Herta Müller, "Every word knows something of a vicious circle," Nobel lecture, December 7, 2009 (https://www .nobelprize.org/prizes/literature/2009/muller/25729-herta -muller-nobel-lecture-2009/).

8 https://www.washingtonpost.com/politics/trump-in-cia -visit-attacks-media-for-coverage-of-his-inaugural-crowds /2017/01/21/f4574dca-e019-11e6-ad42-f3375f271c9c_story .html?utm_term=.a0405f7ed48c

9 Victor Klemperer, *The Language of the Third Reich: LTI: Lingua Tertii Imperii*, trans. Martin Brady, London 2000, p. 14.

10 Viola Zhou, "Beijing party boss promises to eradicate online political rumours ahead of key party congress," *South China Morning Post*, September 27, 2017 (http://www.scmp.com/news/china/policies-politics/article/2113041/beijing-party-boss-promises-eradicate-online-political).

11 Geremie R. Barmé, "New China Newspeak," *China Heritage Quarterly*, No. 29, March 2012 (http://www.chinaheritagequarterly.org/glossary.php?searchterm=029_xinhua.inc&issue=029).

12 Ibid.

13 George Orwell, *1984*, New York, 1949, p. 34.

14 Anna Sun, "The diseased language of Mo Yan," *The Kenyon Review*, autumn 2012 (https://www.kenyonreview.org/kr-online-issue/2012-fall/selections/anna-sun-656342/).

15 Geremie R. Barmé, "New China Newspeak," *China Heritage Quarterly*, No. 29, March 2012 (http://www.chinaheritagequarterly.org/glossary.php?searchterm=029_xinhua.inc&issue=029).

16 Quoted in http://chinaheritage.net/journal/on-new-china-newspeak/

17 Or as other translations have it: "What the superior man requires is just that in his words there may be nothing incorrect." https://china.usc.edu/confucius-analects-13.

18 Mao said this for the first time on August 7, 1927, at an

emergency meeting of the Party's central committee in Wuhan. See: Shao Jianwu, "Mao Zedong qiang ganzi limian chu zhengquan de lishi yu yanbian" ("The story behind Mao Zedong's 'Political power grows out of the barrel of a gun'"), June 13, 2017 (http://dangshi.people .com.cn/n1/2017/0613/c85037-29335727.html).

19 Adrian Zenz, "Domestic security spending: An analysis of available data," *China Brief,* Volume 18, Issue 4, The Jamestown Foundation, March 12, 2018 (https://james town.org/program/chinas-domestic-security-spending -analysis-available-data/).

20 @PDChina, "10,000 pigeons go through anal security check for suspicious objects," *People's Daily* on Twitter, September 30, 2014, 10:48 PM.

21 "Tiaolou shangdiao zhuanghuoche, guanyuan eihe zheme jiduan?" ("They jump to their deaths, they hang themselves, they throw themselves in front of trains: why do some officials commit desperate acts?"), *Beijing News,* April 7, 2017 (http://news.sina.com.cn/c/nd/2017-04-08 /doc-ifyeceza1579240.shtml).

22 Eva Pils et al., "Rule by Fear? A ChinaFile Conversation," ChinaFile, February 18, 2016 (http://www.chinafile.com /conversation/rule-fear).

23 Safeguard Defenders: *Scripted and staged: Behind the scenes of China's forced TV confessions,* CreateSpace Independent Publishing, April 2018.

24 Ibid., p. 24.

25 Jiang Lei, "Zui gao jian: xiang ba da sifa wanzheng li-

angjian" ("Declaring war on the eight sicknesses of the justice system"), *Qiushi*, February 15, 2015 (http://www .qstheory.cn/politics/2015–02/15/c_1114374295.htm).

26 "Cataloging the Torture of Lawyers in China," China Change.org, July 5, 2015 (https://chinachange.org/2015 /07/05/cataloging-the-torture-of-lawyers-in-china/).

27 "Jue bu yunxu 'dang da haishi fa da' weimingti ganrao zhengzhi dingli" ("It is absolutely forbidden to incite political unrest with the misleading question 'Is the Party above the law or the law above the Party?'"), Xinhua, February 5, 2015 (http://www.xinhuanet.com/politics/2015 –02/05/c_1114272511.htm).

28 Andrea Chen,"Rule of law debate pointless in China, *People's Daily* says," *South China Morning Post*, October 29, 2014 (http://www.scmp.com/article/1627657/rule -law-debate-pointless-china-peoples-daily-says).

29 Zhang Ziyang, "Zhou Qiang: Yao ganyu xiang xifang 'sifa duli' deng cuowu sixiang liangjian" ("Chief judge Zhou Qiang: 'We must have the courage to draw our swords against misguided Western ideologies like the independence of the judiciary'"), Chinanews, January 14, 2017 (https://www.thepaper.cn/newsDetail_forward_1600659).

30 Shang Yang, *The Book of Lord Shang*, tr. Yuri Pines, Columbia University Press, 2019, 20.1.

31 Ibid., 18.2.

32 Haifeng Huang, "The Pathology of Hard Propaganda," *Journal of Politics*, December 18, 2017 (https://ssrn.com /abstract=3055019).

33 Sun Liping, "Sixiang shi weihe bei kongzhide" ("Why are thoughts controlled?") (http://www.sohu.com/a/22019 7971_155133).

34 Anthony Kuhn, "China's Few Investigative Journalists Face Increasing Challenges," National Public Radio, August 6, 2017 (https://www.npr.org/sections/parallels/2017 /08/06/539720397/chinas-few-investigative-journalists-face -increasing-challenges).

35 Lin Ping and Xiao An, "China Bans Hip-Hop and Other 'Sub-Cultures' from State Television," Radio Free Asia, January 26, 2018 (https://www.rfa.org/english/news/china /hiphop-ban-01262018105320.html).

36 Yifu Dong et al., "China's Communist Party Takes (Even More) Control of Media: A ChinaFile Conversation," ChinaFile, April 11, 2018 (http://www.chinafile.com/con versation/chinas-communist-party-takes-even-more -control-of-media).

37 Wu Haiyun, "Why Chinese Filmgoers Don't Buy Hollywood's Values Anymore," Sixth Tone, April 9, 2018 (https://www.sixthtone.com/news/1002055/why-chinese -filmgoers-dont-buy-hollywoods-values-anymore).

38 Lu Yan, "Xuyao lixing keguande kan yulunchang" ("Interview with the director of the magazine *Seeking Truth*: a rational and objective consideration of the place of public opinion"), *The Paper*, March 25, 2015 (https://m.thepaper .cn/newsDetail_forward_1314543).

39 Zhao Yinping, "Wangluo daren Xi Jinping" ("Xi Jinping: the wise man of the internet"), Xinhua, Novem-

ber 17, 2016 (http://www.xinhuanet.com/politics/2016–11 /17/c_1119932744.htm).

40 Annie Palmer, "Facebook, Twitter accuse China of running disinformation campaign against Hong Kong protesters" (https://www.cnbc.com/2019/08/19/twitter-accuses-china -of-running-disinformation-campaign-against-hong-kong -protesters.html).

41 Andy Mok, "Patriotic education needs to improve in Hong Kong," *China Daily*, November 6, 2019 (https://www .chinadaily.com.cn/a/201911/06/WS5dc21c4da310cf3e35 575a4e.html).

42 Xuan Yan, "Bu neng rang suanfa jueding neirong" ("The algorithm cannot be allowed to determine content"), *People's Daily*, October 5, 2017 (http://paper.people.com.cn /rmrb/html/2017–10/05/nw.D110000renmrb_20171005 _4–04.htm).

43 The whole letter in Chinese and English can be found in David Bandurski, "Tech Shame in the New Era," China Media Project, April 11, 2018 (http://chinamediaproject .org/2018/04/11/tech-shame-in-the-new-era/).

44 See Leta Hong-Fincher, *Leftover Women: The Resurgence of Gender Inequality in China*, London 2016.

45 Javier C. Hernández and Zoe Mou, "'Me too,' Chinese Women Say. 'Not So Fast,' Say the Censors," *New York Times*, January 23, 2018 (https://www.nytimes.com/2018 /01/23/world/asia/china-women-me-too-censorship.html).

46 https://chinadigitaltimes.net/2018/04/minitrue-do-not-report-on-peking-university-open-letter/.

47 Gary King, Jennifer Pan, and Margaret E. Roberts, "How the Chinese Government Fabricates Social Media Posts for Strategic Distraction, not Engaged Argument," *American Political Science Review*, 111, 3, April 2017, pp. 484–501.

48 Ibid., p. 499.

49 Kristin Shi-Kupfer, Mareike Kohlberg, Simon Lang, and Bertram Lang, "Ideas and ideologies competing for China's political future," MERICS Papers on China, No. 5, October 2017, p. 10.

50 "Zhulao guojia wangluo anquan pingzhang" ("Building a security shield for the national internet"), *People's Daily*, April 23, 2018 (opinion.people.com.cn/n1/2018/0423/c1003 –29943655.html).

51 David Bandurski, "China's 'Great Firewall' is more akin to a 'Great Hive' of propaganda buzzing around individuals," *Hong Kong Free Press*, September 24, 2017 (https://www.hongkongfp.com/2017/09/24/chinas-great-firewall -akin-great-hive-propaganda-buzzing-around-individuals/).

52 "Institutional Strength: China's Key to Beating Novel Coronavirus," Xin-hua, March 3, 2020, (www.xinhuanet .com/english/2020-03/10/c_138863498.htm).

53 The first English translation of the complete essay can be found here: Ren Zhiqiang, "My reading of February 23rd" (http://credibletarget.net/notes/RZQ).

54 Neil Postman, *Amusing Ourselves to Death: Public Discourse in the Age of Showbusiness*, New York 1985.

55 Aldous Huxley, *Brave New World Revisited*, 1958 (www .huxley.net/bnw-revisited/).

56 Yuyu Chen and David Y. Yang, "The Impact of Media Censorship—Evidence from a Field Experiment in China," Stanford Graduate School of Business, January 4, 2018.

57 Ibid., p. 2.

58 Ibid.

59 *The complete works of Zhuangzi*, tr. Burton Watson, New York 2013, p. 126.

60 Fang Lizhi (translated by Perry Link), "The Chinese Amnesia," *New York Review of Books*, September 27, 1990 (https://www.nybooks.com/articles/1990/09/27/the-chinese -amnesia/).

61 Louisa Lim, *The People's Republic of Amnesia: Tiananmen Revisited*, Oxford 2015.

62 Yan Lianke, "On China's State-Sponsored Amnesia," *New York Times*, April 1, 2013 (https://www.nytimes .com/2013/04/02/opinion/on-chinas-state-sponsored- amnesia.html).

63 Yan Lianke, "What Happens After Coronavirus? On Community Memory and Repeating Our Own Mistakes," Literary Hub, March 11, 2020 (https://lithub.com/yan -lianke-what-happens-after-coronavirus).

64 Yang Jisheng, *Tombstone: the untold story of Mao's great famine*, New York 2012.

65 Frank Dikötter, *Mao's Great Famine: The History of China's Most Devastating Catastrophe, 1958–1962*, London 2010.

66 Perry Link, "Politics and the Chinese Language: What Mo Yan Defenders Get Wrong," Asia Society, Asia Blog,

December 27, 2012 (https://asiasociety.org/blog/asia/pol itics-and-chinese-language-what-mo-yans-defenders-get -wrong).

67 See Deng Xiaoci, "China defends Long March," *Global Times*, September 26, 2016 (http://www.globaltimes.cn /content/1008494.shtml).

68 Zheng Yi, *Scarlet Memorial: Tales of Cannibalism in Modern China*, London 1998.

69 "Wenchuan queli 'gan'en ri', rang ai de yongquan benli-ubuxi" ("Wenchuan proclaims 'day of gratitude', let the bubbling springs flow and never run dry"), Xinhua, May 6, 2018 (www.xinhuanet.com/2018-05/06/c_1122790543.htm).

70 Simon Leys, "The Art of Interpreting Nonexistent Inscriptions Written in Invisible Ink on a Blank Page," *The New York Review of Books*, October 11, 1990 (http:// www.nybooks.com/articles/1990/10/11/the-art-of-inter preting-nonexistent-inscriptions-w/).

71 Richard McGregor, *The Party: The Secret World of China's Communist Rulers*, London 2010.

72 Speech by Xi Jinping at the 19th Party congress of the CCP on October 18, 2017.

73 "Zhonggong zhongyang yinfa 'shenhua dang he guo-jia jigou gaige fang'an'" (Central committee of the CCP: "Plan for reform of the Party and state organs"), Xinhua, March 21, 2018 (http://www.xinhuanet.com/pol itics/2018–03/21/c_1122570517.htm).

74 Alexandra Stevenson, "China's Communists Rewrite the Rules for Foreign Businesses," *New York Times*, April 13,

2018 (https://www.nytimes.com/2018/04/13/business/china-communist-party-foreign-businesses.html).

75 Yu Mengtong, "Zhongguo fayuan yuanzhang: 'dangxing renxing chongtu shi jianchi dangxing'" ("Chinese chief judge: 'In a conflict between Party nature and human nature, one must adhere to the Party nature'"), May 17, 2016 (https://www.voachinese.com/a/chinese-judge-says-party-principle-before-human-nature-20160516/3332419.html).

76 Christopher Balding, Twitter post, July 26, 2017 (https://twitter.com/BaldingsWorld/status/890265518124310528).

77 David Bandursky, "Building the Party's Internet," China Media Project, May 11, 2018 (chinamediaproject.org/2018/05/11/building-the-partys-internet/).

78 Ibid.

79 Peng Lihui, "Ma Yun: 'Xianjin zhongguo shi zui jia jingshang shidai, buyong gao momingqimiao guanxi'" ("Ma Yun: 'Doing business has never been so good as it is in China today, you no longer need to rely on your connections'"), November 29, 2017 (http://tech.163.com/17/1129/23/D4EP3UDD00097U7R.html).

80 "Liang Wengen, 'Wode caichan jizhi shengming dou shi dangde'" ("My property and even my life belong to the party"), Guangming.com (https://news.ifeng.com/mainland/special/zhonggong18da/detail_2012_11/12/19047721_0.shtml).

81 Li Hua, "Ningbo bianyin dangyuan ganbu fumian yanxing tixingben, liechu 68 xiang" ("Ningbo brings out a

handbook to remind people of negative words and deeds of Party members and lists 68 examples"), *Qianjiang Abendnachrichten,* June 24, 2016 (http://ningbo.news.163 .com/16/0624/11/BQAQE3FD03431AF0.html).

82 Geremie R. Barmé, "For Truly Great Men, Look to This Age Alone," *China Heritage,* January 27, 2018 (http:// chinaheritage.net/journal/for-truly-great-men-look-to -this-age-alone/).

83 Chris Udemans, "Chinese propaganda app puts user data at risk: researchers," in "With Chinese Characteristics," October 14, 2019.

84 "Stick to Karl Marx's true path, Xi Jinping tells China's communists," *South China Morning Post,* May 4, 2018 (www.scmp.com/news/china/policies-politics/article/2144 716/stick-karl-marxs-true-path-xi-jinping-tells-chinas #link_time=1525427622).

85 Zhou Yu, "Chinese university revives research on official ideology to head off suspicious values," *Global Times,* June 1, 2015 (http://www.globaltimes.cn/content/924726 .shtml).

86 Li Rohan, "Students 'inspired' by ideology course: poll," *Global Times,* March 16, 2018 (http://www.globaltimes .cn/content/1093710.shtml).

87 Rong Jian, "A China Bereft of Thought," *The China Story,* February 5, 2017 (https://www.thechinastory.org /cot/rong-jian-%E8%8D%A3%E5%89%91-on-thought -and-scholarship-in-china/).

88 "Leaked Speech Shows Xi Jinping's Opposition to Re-

form," *China Digital Times*, January 27, 2013 (https://chinadigitaltimes.net/2013/01/leaked-speech-shows-xi-jinpings-opposition-to-reform/).

89 Oiwan Lam, "The Chinese Communist Party Forbids Members from Celebrating Christmas, Calling It a Festival of Humiliation," *Global Voices*, December 24, 2017 (https://globalvoices.org/2017/12/24/the-chinese-communist-party-forbids-members-from-celebrating-christmas-calling-it-a-festival-of-humiliation/).

90 "Xi Jinping zhuxi Qufu jianghua: shijie ruxue chuanbo, zhongguo yao baochi chongfen huayuquan" ("The speech by Chairman Xi Jinping in Qufu: on the spread of Confucianism in the world, China still has the final word"), *Guancha*, September 29, 2014 (http://www.guancha.cn/XiJinPing/2014_09_29_271934.shtml).

91 English translation of this speech: http://library.chinausfocus.com/article-1534.html.

92 Li Ling, "Marxism, the CCP, and Traditional Chinese Culture as I Know Them," *China Digital Times*, July 5, 2017 (https://chinadigitaltimes.net/2017/07/li-ling-marxism-ccp-traditional-chinese-culture/).

93 State Council of the People's Republic of China and Office of the Central Committee of the CCP: "Guanyu shixuan zhonghua youxiu chuantong wenhua chuancheng fazhan gongcheng de yijian" ("Implementation of the project of spreading and developing the great traditions of Chinese culture") (http://www.gov.cn/zhengce/2017–01/25/content_5163472.htm).

94 Ernest Gellner, *Nations and Nationalism*, New York 1983.

95 Liu Mingfu, *The China Dream*, Beijing 2015.

96 https://www.newyorker.com/magazine/2017/06/19/are-china-and-the-united-states-headed-for-war.

97 "Zuguo shi ni weinanshi de houzhi, dan bu shi beiguoxia" ("The fatherland stands behind you in a crisis, but is not your scapegoat"), CCTV online, February 2, 2018 (http://news.cctv.com/2018/02/02/ARTIKRDxg8O3R8cePwGTy UIb180202.shtml).

98 "China a positive force for world peace," Xinhua, August 1, 2017 (http://www.xinhuanet.com/english/2017-08/01/c_136490347.htm).

99 Jeremy Bentham (1843d), *The Works*, 4, Panopticon, Constitution, Colonies, Codification, Liberty fund. p. 39.

100 Jeremy Bentham (1843d), *The Works*, 10, Memoirs Part I and Correspondence, Liberty fund.

101 Michel Foucault, *Discipline and Punish: The birth of the prison*, New York 1977 (French original published in 1975).

102 Jason Dean, "Why artificial intelligence is the new electricity," *Wall Street Journal*, July 27, 2017 (https://www.marketwatch.com/story/why-artificial-intelligence-is-the-new-electricity-2017-06-27).

103 Jeffrey Ding, "Deciphering China's AI Dream—The context, components, capabilities and consequences of China's strategy to lead the world in AI," Future of Humanity Institute, University of Oxford, March 2018, p. 7.

104 "Xi Jinping: Tuidong woguo xinyidai rengong zhineng jiankang fazhan" ("Xi Jinping: driving forward the healthy development of a new generation of artificial intelligence"), in Xinhua, October 31, 2018 (http://cpc.people .com.cn/n1/2018/1031/c64094-30374719.html).

105 "State Council Notice on the Issuance of the Next Generation Artificial Intelligence Development Plan," translated by China Copyright and Media, July 20, 2017 (https:// chinacopyrightandmedia.wordpress.com/2017/07/20/a -next-generation-artificial-intelligence-development-plan/).

106 Zhi Zhenfeng, "Jiakuaituijin wangluoqiangguo sianshe de genben zunxun" ("Fundamentals for expediting and constructing a cyber-power"), *Guangming Daily*, May 7, 2018 (epaper.gmw.cn/gmrb/html/2018–05/07/nw.D110000 gmrb_20180507_1_11.htm).

107 "White Paper Outlines Potential Uses of AI," *China Digital Times*, November 19, 2018, (https://chinadigitaltimes .net/2018/11/white-paper-outlines-potential-uses-of -artificial-intelligence/).

108 https://homes.cs.washington.edu/~pedrod/Prologue.pdf.

109 Emily Rauhala, "America wants to believe China can't innovate. Tech tells a different story," in *Washington Post*, July 19, 2016 (https://www.washingtonpost.com/world /asia_pacific/americawants-to-believe-china-cant-inno vate-tech-tells-a-different-story/2016/07/19/c17cbea9-6e e6-479c-81fa-54051df598c5_story.html?noredirect=on &utm_term=.4f67f508b2de).

110 National Science Board, "Science & Engineering Indicators 2018," National Science Foundation, January 2018 (https://www.nsf.gov/statistics/2018/nsb20181/).

111 "Xi stresses development, application of blockchain technology," Xinhua, October 25, 2019 (http://www.xinhuanet.com/english/2019–10/25/c_138503254.htm).

112 "Beijing fayuan: dangshiren ke tongguo weixin li'an" ("Beijing court: trial participants can submit files using Weixin"), Xinhua, December 26, 2017 (http://www.xinhuanet.com/legal/2017–12/26/c_1122168318.htm).

113 "Guangzhou qianfa quanguo shouzhang weixin shenfenzheng" ("Guangzhou issues first WeChat ID cards"), Xinhua, December 27, 2017 (http://www.xinhuanet.com/city/2017–12/26/c_129775203.htm).

114 Jeffrey Ding, see note 103, above, p. 25.

115 Ibid.

116 Zhou Wenting, "AI takes a look at legal evidence," *China Daily*, July 11, 2017 (http://www.chinadaily.com.cn/china/2017–07/11/content_30064693.htm).

117 "China embraces AI: A close look and a long view," Sinovation Ventures/Eurasia Group, 2017 (https://www.eurasiagroup.net/files/upload/China_Embraces_AI.pdf).

118 "State Council Notice on the Issuance of the Next Generation Artificial Intelligence Development Plan," as 105, above.

119 July 27, 2019, 06:30 来源: 澎湃新闻·澎湃号·政务, THE PAPER.

120 https://www.scmp.com/tech/enterprises/article/2126553

/metro-pickpockets-beware-chinas-hi-tech-cameras-are
-watching-your

121 Sina News: "Hangzhou zhe suo zhongxue jiaoshi limi-
ande 'Tianyan'" ("The 'eyes in the sky' in the classrooms
of this middle school in Hangzhou"), May 16, 2018 (slide
.news.sina.com.cn/s/slide_1_2841_271359.html#p=7).

122 Paul Bischoff, "The world's most-surveilled cities," Comparitech, August 15, 2019 (https://www.comparitech.com
/vpn-privacy/the-worlds-most-surveilled-cities/).

123 Human Rights Watch: "China: Police 'Big Data' Systems Violate Privacy, Target Dissent," November 19, 2017
(https://www.hrw.org/news/2017/11/19/china-police-big
-data-systems-violate-privacy-target-dissent).

124 https://www.hrw.org/news/2017/11/19/china-police-big
-data-systems-violate-privacy-target-dissen

125 Tang Yu, "Dashuju shidai 'hulianwang + jingwu' shengji
shehuizhili moshi" ("The era of big data: how internet
and police together are taking the model of social management to a new level"), *Newspaper for Democracy
and Law*, January 4, 2017 (http://www.cbdio.com/Big
Data/2017–01/04/content_5422185.htm).

126 Wang Yongqing, "Wanshan shehui zhian zonghe zhili
tizhi jizhi" ("Perfecting the mechanisms of the system
of comprehensive steering of public security"), *Qiushi*,
2015/22 (http://www.qstheory.cn/dukan/qs/2015–11/15/c
_1117135295.htm).

127 Wang Jingshan, "Dongyingshi gonganju jingxin dazao
'zaobadian + jingwu' dafangguankong xin moshi" ("The

Dongying police and their carefully implemented new model of crime-fighting and prevention, with the help of the 'Mornings at Eight' police project"), *Shandong Legal Daily*, August 17, 2016 (http://fazhi.dzwww.com /dj/201608/t20160817_10786479.htm).

128 Charles Rollet, "Hikvision Markets Uyghur Ethnicity Analytics, Now Covers Up," IPVM, November 11, 2019 (https://ipvm.com/reports/hikvision-uyghur).

129 Paul Monzur, "One Month, 500,000 Face Scans: How China Is Using A.I. to Profile a Minority," *New York Times*, April 14, 2019 (https://www.nytimes.com/2019 /04/14/technology/china-surveillance-artificial-intelligence -racial-profiling.html).

130 Sijia Jiang, "Huawei founder details 'battle mode' reform plan to beat U.S. crisis," Reuters, August 20, 2019 (https://www.reuters.com/article/us-huawei-tech-founder /huawei-founder-details-battle-mode-reform-plan-to-beat -u-s-crisis-idUSKCN1VA0Z0).

131 Kurban Niyaz, "Authorities Require Uyghurs in Xinjiang's Aksu to Get Barcodes on Their Knives," Radio Free Asia, October 11, 2017 (https://www.rfa.org/english/news /uyghur/authorities-require-uyghurs-in-xinjiangs-aksu-to -get-barcodes-on-their-knives-10112017143950.html).

132 Austin Ramzy and Chris Buckley, " 'Absolutely No Mercy': Leaked Files Expose How China Organized Mass Detentions of Muslims," *New York Times*, November 16, 2019 (https://www.nytimes.com/interactive

/2019/11/16/world/asia/china-xinjiang-documents.html?
action=click&module=RelatedLinks&pgtype=Article).

133 "China's Algorithms of Repression: Reverse Engineering a Xinjiang Police Mass Surveillance App," Human
Rights Watch, August 7, 2019 (https://www.hrw.org/
report/2019/05/01/chinas-algorithms-repression/reverse-
engineering-xinjiang-police-mass-surveillance).

134 Shai Oster, "China Tries Its Hand at Pre-Crime," Bloomberg, March 4, 2016 (https://www.bloomberg.com/news
/articles/2016-03-03/china-tries-its-hand-at-pre-crime).

135 "Siemens seals strategic cooperation agreement with
China Electronics Technology Group Corporation," Siemens Press Release. (https://w1.siemens.com.cn/news_en
/news_articles_en/6780.aspx).

136 "China Cables. Exposed: China's Operating Manuals
for Mass Internment and Arrest by Algorithm," International Consortium of Investigative Journalists ICIJ, November 24, 2019 (https://www.icij.org/investigations/china
-cables/exposed-chinas-operating-manuals-for-mass-
internment-and-arrest-by-algorithm/).

137 "At Least 150 Detainees Have Died in One Xinjiang Internment Camp: Police Officer," Radio Free Asia, October 29, 2019 (https://www.rfa.org/english/news/uyghur
/deaths-10292019181322.html).

138 https://www.icij.org/investigations/china-cables/read-the
-china-cables-documents/.

139 Adrian Zenz, " 'Wash Brains, Cleanse Hearts': Evidence

from Chinese Government Documents about the Nature and Extent of Xinjiang's Extrajudicial Internment Campaign," *Journal of Political Risk*, Vol. 7, No. 11, November 2019 (http://www.jpolrisk.com/wash-brains-cleanse -hearts/).

140 Adrian Zenz, "New Evidence for China's Political Re-Education Campaign in Xinjiang," Jamestown Foundation, May 15, 2018 (https://jamestown.org/program /evidence-for-chinas-political-re-education-campaign-in -xinjiang).

141 "'Pingan Xinjiang zhihui tongxing' lianhe chuangxin shiyanshi luohu wulumuqi gaoxinqu" ("A new research laboratory called 'Peaceful Xinjiang, Smart Cooperation' has been opened in the Gaoxin district of Urumqi"), Urumqi High-Tech Industrial Development Zone, May 11, 2018 (www.uhdz.gov.con/info/1005/25748.htm).

142 Danielle Cave, Fergus Ryan, and Vicky Xiuzhong Xu, "Mapping more of China's tech giants: AI and surveillance," Australian Strategic Policy Institute, November 28, 2019 (https://www.aspi.org.au/report/mapping -more-chinas-tech-giants).

143 Emily Feng, "Uighur children fall victim to China anti-terror drive," *Financial Times*, July 10, 2018 (https:// www.ft.com/content/f0d3223a-7f4d-11e8-bc55-50daf11b7 20d).

144 Tencent Research Institute (ed.), *Rengongzhineng: Guojia rengongzhineng zhanlüe xingdong zhuashou* ("Artificial Intelligence: a national strategic initiative for

AI"), Renmin Daxue Chubanshe (People's University of China Press), November 2017.

145 "Zhongguoren geng kaifang, yuanyong yinsi huan xiaolü" ("The Chinese are more open, they are willing to exchange their privacy for efficiency"), *The Paper*, March 26, 2018 (http://tech.caijing.com.cn/20180326/4425 045.shtml).

146 Sun Qiru, "Rengong zhineng weixie geren yinsi cheng yinyou" ("Secret concerns about the threat to the privacy from artificial intelligence"), *Beijing News*, March 3, 2018 (http://www.xinhuanet.com/info/2018–03/03/c_1370 12166.htm).

147 Jiang Jie, "China's leading facial recognition beefs up nation's surveillance network for security," *People's Daily Online*, November 20, 2017 (http://en.people.cn/n3/2017 /1120/c90000-9294587.html).

148 State Council of the People's Republic of China, "Planning Outline for the Establishment of a Social Credit System (2014–2020)," June 14, 2014 (https://www.chinalaw translate.com/socialcreditsystem/?lang=en).

149 Credit China, "Laolai xianqin jianguangsi mo yin shixin wu zhongsheng" ("An incorrigible debtor being exposed on a date. Don't ruin your whole life with your breaches of trust"), November 14, 2017 (https://www.creditchina .gov.cn/gonggongwenjianjia/tujiexinyongold/tujiexinyong /201711/t20171114_93948.html).

150 Sinolytics, "The Digital Hand. How China's Corporate Social Credit System Conditions Market Actors," Au-

gust 28, 2019 (https://www.europeanchamber.com.cn/en
/press-releases/3045/european_chamber_report_on_china
_s_corporate_social_credit_system_a_wake_up_call_for
_european_business_in_china).

151 Gao Feng, "Beijing Metro to Begin Security 'Sorting'
Based on Facial Recognition" October 30, 2019 (https://
www.rfa.org/english/news/china/facial-10302019162222
.html).

152 Mareike Ohlberg, Shazeda Ahmed, and Bertram Lang,
"Central Planning, Local Experiments: The complex implementation of China's Social Credit System," MERICS
China Monitor, December 12, 2017 (https://www.merics
.org/sites/default/files/2017–12/171212_China_Monitor
_43_Social_Credit_System_Implementation.pdf).

153 Andreas Landwehr, "Brisantes Handbuch: China schnürt
ausländischen NGOs die Luft ab," dpa, May 6, 2018
(www.nzz.ch/international/brisantes-handbuch-schnuert
-auslaendischen-ngosdie-luft-ab-ld.1383554).

154 Interview with Sebastian Heilmann, "19. Parteitag: Die
Digitalisierung spielt der Kommunistischen Partei in die
Hände," MERICS China Flash, October 11, 2017 (https://
www.merics.org/de/chinaflash/19-parteitag-der-kp
-china-die-digitalisierung-spielt-der-kommunistischen
-partei-die).

155 Yan Lianke, "Finding Light in China's Darkness," New
York Times, October 22, 2014 (https://www.nytimes
.com/2014/10/23/opinion/Yan-Lianke-finding-light-in
-chinas-darkness.html).

156 "What worries the world?" Ipsos Public Affairs, July 2017 (https://www.ipsos.com/en/what-worries-world-july-2017).

157 Hannah Arendt, *The Origins of Totalitarianism*, London 2017, p. XI.

158 Nancy Bazilchuck, "Chinese are more right wing than Americans," *Science Nordic*, October 8, 2018 (http://sci encenordic.com/chinese-are-more-right-wing-americans).

159 Cheng Yinghong, "Diduan renkou: shehuidaerwenzhuyi zhengzhi de buxiangzhizhao" ("The underclass—harbingers of Social Darwinism"), *The Intitium*, November 26, 2017 (https://theinitium.com/article/20171126 -opinion-Social-Darwinism/).

160 Murong Xuecun: "Ru qiushui changtian" ("Autumn water and endless sky"), speech at the Hong Kong Book Fair, July 25, 2012. English translation: "A Plea for a gentler China," *China Digital Times*, July 30, 2012 (https:// chinadigitaltimes.net/2012/07/one-authors-plea-for-a -gentler-china/).

161 Sebastian Herrmann, "Warum das Vorbild nicht zu vor-bildlich sein darf," *Süddeutsche Zeitung*, March 3, 2018 (http://www.sueddeutsche.de/wissen/psychologie-warum -das-vorbild-nicht-zuvorbildlich-sein-darf-1.3402683).

162 Chang Ping, "We'd be satisfied with any government," ChinaChange.org, October 1, 2015 (https://chinachange .org/2015/10/01/wed-be-satisfied-with-any-government/).

163 Ai Weiwei, "Ni ruguo xiwang liaojie nide zuguo, ni yijing zoushangle fanzui de daolu," Tweet by Ai Weiwei, March 30, 2010.

164 Ra Jong Yil, "The darkness of heart," May 15, 2012 (http://destinationpyongyang.blogspot.com/2012/05/darkness-of-heart-by-ra-jong-yil.html).

165 http://shanghaiist.com/2015/08/28/wang_sicong_secrets_of_china/

166 Weibo user Yuanliuqingnian, "Normal Country Delusion," *China Digital Times*, August 18, 2015 (https://chinadigitaltimes.net/2015/08/translation-normal-country-delusion-in-tianjin/).

167 Minxin Pei, "Chinese history tells us: Never stop fighting until the fight is done," *Nikkei Asian Review*, July 27, 2017 (https://asia.nikkei.com/Politics/Chinese-history-tells-us-Never-stop-fighting-till-the-fight-is-done).

168 Li Chenjian, "Tingzhi beiliang, juzuo quanru" ("Stand tall and deny yourselves cynicism"), *Mingpao*, March 24, 2018 (https://news.mingpao.com/ins/instantnews/web_tc/article/20180324/s00004/1521904316201).

169 Li Ling, "Marxism, the CCP, and Traditional Chinese Culture as I Know Them," *China Digital Times*, July 5, 2017 (https://chinadigitaltimes.net/2017/07/li-ling-marxism-ccp-traditional-chineseculture/).

170 Karin Betz, "Schreib mir mit deiner Asche einen Brief," Interview with Liao Yiwu, *Neue Zürcher Zeitung*, July 22, 2017 (https://www.nzz.ch/feuilleton/liao-yiwu-ueber-den-tod-des-buerger-rechtlers-liu-xiaobo-schreib-mir-mit-deiner-asche-einen-brief-ld.1307297).

171 Geremie R. Barmé, "The True story of Lu Xun," *The*

New York Review of Books, November 23, 2017 (http://www.nybooks.com/articles/2017/11/23/true-story-of-lu-xun/).

172 Zhao Xiaoli, "If I don't speak, I'll spend my life in shame," *China Digital Times,* February 25, 2018 (https://chinadigitaltimes.net/2018/02/translation-dont-speak-ill-spend-life-shame/).

173 Václav Havel, "The Power of the Powerless," tr. Paul Wilson, *The Power of the Powerless: Citizens Against the State in Central-Eastern Europe,* London 2009, p. 23.

174 English translation: Xiao Meili, "The Story of Ma Hu," *China Digital Times,* June 20, 2018 (https://chinadigitaltimes.net/2018/06/translation-the-story-of-ma-hu-part-3/).

175 Benjamin Peters, *How Not to Network a Nation: The Uneasy History of The Soviet Internet,* Cambridge, MA, 2016.

176 Preface to the *Novissima Sinica* (1697/1699), http://east_west_dialogue.tripod.com/id12.html.

177 Ian Buruma, "Year of the 'China model,'" *The Guardian,* January 9, 2008 (https://www.theguardian.com/commentisfree/2008/jan/09/yearofthechinamodel).

178 Michael Moritz, "China is leaving Donald Trump's America behind," *Financial Times,* September 10, 2017 (https://www.ft.com/content/1ac0337c-9470–11e7–83ab-f4624cccbabe).

179 Michael Moritz, "Silicon Valley would be wise to fol-

low China's lead," *Financial Times*, January 18, 2018 (https://www.ft.com/content/42daca9e-facc-11e7–9bfc-052cbba03425).

180 https://ec.europa.eu/commission/sites/beta-political/files/communication-eu-china-a-strategic-outlook.pdf.

181 https://www.theguardian.com/stage/2018/apr/04/unseen-letters-shed-light-on-royal-court-censorship-row-british-council.

182 Dan Strumpf, "Apple's Tim Cook: No Point Yelling At China," *Wall Street Journal*, December 7, 2017 (https://www.wsj.com/articles/apples-tim-cook-no-point-yelling-at-china-1512563332).

183 Yuan Yang, "Microsoft worked with Chinese military university on artificial intelligence," *Financial Times*, April 10, 2019, (https://www.ft.com/content/9378e7ee-5ae6–11e9–9dde-7aedca0a081a).

184 https://www.blog.google/around-the-globe/google-asia/google-ai-china-center.

185 Ryan Gallagher, "Google Employees Uncover Ongoing Work on Censored China Search," *The Intercept*, March 4, 2019 (https://theintercept.com/2019/03/04/google-ongoing-project-dragonfly/).

186 "Xi stresses modernizing China's governance system, capacity," Xinhua, September 25, 2019 (http://www.xinhuanet.com/english/2019–09/25/c_138419199.htm).

187 Li Laifang, "Enlightened Chinese democracy puts the West in the Shade," Xinhua, October 17, 2017 (http://

www.xinhuanet.com/english/2017–10/17/c_136685546
.htm).

188 Teddy Ng, Andrea Chen, "Xi Jinping says world has
nothing to fear from awakening of 'peaceful lion,'" *South
China Morning Post*, March 28, 2014 (http://www.scmp
.com/news/china/article/1459168/xi-says-world-has
-nothing-fear-awakening-peaceful-lion).

189 See for example Song Yanbin, "China's Foreign Policy,"
speech by Song Yanbin, ambassador of the PRC to Costa
Rica, at the UN University for Peace, Ciudad Colón
(Costa Rica), December 2016.

190 https://twitter.com/betoorourke/status/11810470928751575
04?lang=en.

191 Thorsten Benner, Jan Gaspers, Mareike Ohlberg, Lucre-
zia Poggetti, and Kristin Shi-Kupfer, "Authoritarian Advance: Responding to China's Growing Political Influence
in Europe," Berlin: MERICS and GPPI, February 2018
(http://www.gppi.net/publications/rising-powers/article
/authoritarian-advance-responding-to-chinas-growing
political-influence-in-europe/).

192 https://www.cecc.gov/events/hearings/the-long-arm-of
-china-exporting-authoritarianism-with-chinese-charac
teristics.

193 "Power and Influence: The hard edge of China's Soft
Power," June 5, 2017, abc.net.au (http://www.abc.net.au
/4corners/power-and-influence-promo/8579844).

194 https://www.smh.com.au/opinion/the-real-reason-you

-wont-be-reading-my-new-book-on-china-anytime-soon
-20171124-gzsjse.html.

195 Clive Hamilton, *Silent Invasion: China's Influence in Australia*, Melbourne 2018.

196 "CENSORSHIP AND CONSCIENCE: FOREIGN AUTHORS AND THE CHALLENGE OF CHINESE CENSORSHIP," PEN American Center, May 20, 2015, (https://pen.org/sites/default/files/PEN%20Censorship %20and%20Conscience%202%20June.pdf).

197 "Beijing Censors Foreign Books Printed in China," China Digital Times, February 26, 2019 (https://china digitaltimes.net/2019/02/beijing-censors-foreign-books -printed-in-china/).

198 Anne-Marie Brady: *Magic Weapons: China's political influence activities under Xi Jinping*, Christchurch, September 2017.

199 https://www.thetimes.co.uk/article/new-zealand-aca demic-anne-marie-brady-is-targeted-for-exposing-china -885gkh5vq.

200 Eleanor Ainge Roy, " 'I'm being watched': Anne-Marie Brady, the China critic living in fear of Beijing," *The Guardian*, January 23, 2019 (https://www.theguardian .com/world/2019/jan/23/im-being-watched-anne-marie -brady-the-china-critic-living-in-fear-of-beijing).

201 George Gao, "Why is China so . . . uncool?," *Foreign Policy* March 8, 2017 (http://foreignpolicy.com/2017/03/08 /why-is-china-so-uncool-soft-power-beijing-censorship -generation-gap/).

202 David Shambaugh, "China's Soft Power Push," *Foreign Affairs*, July/August2015 (https://www.foreignaffairs.com /articles/china/2015-06-16/china-s-soft-power-push).

203 Anne-Marie Brady, *Magic Weapons: China's political influence activities under Xi Jinping*, Christchurch, September 2017.

204 James King, Lucy Hornby, and Jamie Anderlei, "Inside China's secret 'magic weapon' for worldwide influence," *Financial Times*, October 26, 2017 (https://www.ft.com /content/fb2b3934-b004-11e7-beba-5521c713abf4).

205 "Dalai Lama can't deny China central government's role in reincarnation," Xinhua, September 6, 2015 (http:// www.xinhuanet.com/english/video/2015–09/07/c_1345 98280.htm).

206 Bethany Allen-Ebrahimian, "The Moral Hazard of Dealing With China," *The Atlantic*, January 11, 2020 (https:// www.theatlantic.com/international/archive/2020/01/ste phen-schwarzman-china-surveillance-scholars-colleges /604675/).

207 Charles Parton, "China-UK Relations. Where to Draw the Border Between Influence and Interference," Royal United Services Institute, RUSI Occasional Paper, February 2019 (https://rusi.org/event/china%E2%80%93uk-relations -where-draw-border-between-influence-and-interference).

208 Wang Dan, "Beijing Hinders Free Speech in America," *New York Times*, November 26, 2017 (https://www .nytimes.com/2017/11/26/opinion/beijing-free-speech -america.html).

209 Bethany Allen-Ebrahimian, "The Chinese Communist Party is Setting Up Cells at Universities in America," *Foreign Policy*, April 18, 2018 (http://foreignpolicy .com/2018/04/18/the-chinese-communist-party-is-set ting-up-cells-at-universities-across-america-china-students-beijing-surveillance/).

210 Bethany Allen-Ebrahimian, "China's Long Arm Reaches into American Campuses," *Foreign Policy*, March 7, 2018 (http://foreignpolicy.com/2018/03/07/chinas-long-arm -reaches-into-american-campuses-chinese-students-scholars-association-university-communist-party/).

211 "Rang zhonghua ernü gongxiang xingfu guangrong" ("Let the sons and daughters of China share in happiness and glory"), *People's Daily*, March 21, 2018 (http:// lianghui.people.com.cn/2018npc/n1/2018/0320/c417507 –29879275.html).

212 https://twitter.com/yangyang_cheng/status/9653518745 71505665.

213 Kinas ambassadör, "Vi har hagelgevär för våra fiender," Expressen, November 30, 2019 (https://www.expressen .se/nyheter/kinas-ambassador-vi-har-hagelgevar-for-vara -fiender/).

214 Bethany Allen-Ebrahimian, "Chinese Police Are Demanding Personal Information from Uighurs in France," in *Foreign Policy*, March 2, 2018 (http://foreignpolicy .com/2018/03/02/chinese-police-are-secretly-demanding -personal-information-from-french-citizens-uighurs-xin jiang/).

215 "A tour of Confucius Institutes in the Americas," *China Daily*, May 30, 2014 (http://global.chinadaily.com.cn/a /201405/30/WS5a2fc6f3a3108bc8c6729bfa.html).

216 Continually updated list of Confucius Institutes in the United States: https://www.nas.org/blogs/dicta/how_many _confucius_institutes_are_in_the_united_states.

217 http://chinesestudies.eu/?p=585.

218 "Confucius Institutes to better serve Chinese diplomacy," *People's Daily Online*, January 25, 2018 (http://en.people .cn/n3/2018/0125/c90000–9419574.html).

219 https://www.cambridge.org/about-us/media/press-re leases/cambridge-university-press-statement-regarding -content-china-quarterly.

220 https://twitter.com/Rory_Medcalf/status/8986718443 91231488.

221 https://mobile.twitter.com/gregdistelhorst/status/8985 86072346615808.

222 https://chinadigitaltimes.net/2017/08/cambridge- university-press-makes-u-turn-chinese-censorship/.

223 https://www.cambridge.org/about-us/media/press-re leases/cambridge-university-press-statement-regarding -content-china-quarterly.

224 Charles Parton, "China-UK Relations. Where to Draw the Border Between Influence and Interference," Royal United Services Institute, RUSI Occasional Paper, February 2019 (https://rusi.org/publication/occasional-papers /china-uk-relations-where-draw-border-between-influ ence-and).

225 Jim Pickard, "Cameron faces scrutiny on UK-China fund," *Financial Times*, April 22, 2018 (https://www.ft.com/content/b558418a-43ee-11e8-803a-295c97e6fd0b).

226 Harold Thibault and Brice Pedroletti, "Les tribulations de Jean-Pierre Raffarin en Chine," *Le Monde*, December 6, 2019 (https://www.lemonde.fr/international/article/2019/12/06/les-tribulations-de-jean-pierre-raffarin-en-chine_6021867_3210.html).

227 Yuichiro Kakutani, "China Violates Disclosure Law to Publish Propaganda in NY Times, WaPo," *Washington Free Beacon*, December 18, 2019 (https://freebeacon.com/national-security/china-flouts-fed-law-to-publish-propaganda-in-ny-times-wapo/).

228 Koh Gui Qing and John Shiffman, "Exposed: China's covert global radio network," Reuters, November 2, 2015 (https://www.reuters.com/investigates/special-report/china-radio/).

229 Rebecca Davis, "China Aims to Become 'Strong Film Power' Like U.S. by 2035, Calls for More Patriotic Films," *Variety*, March 3, 2019 (https://variety.com/2019/film/news/china-strong-film-power-by-2035-wants-more-patriotic-films-1203153901/).

230 May Jeong, " 'The Big Error Was That She Was Caught': The Untold Story Behind the Mysterious Disappearance of Fan Bingbing, the World's Biggest Movie Star," *Variety*, March 26, 2019.

231 http://undocs.org/A/HRC/37/L.36.

232 Lin Songtian, "China is setting up a new model for world human rights," *People's Daily Online*, January 17, 2019 (http://en.people.cn/n3/2019/0117/c90000-9538911.html).

233 Thailand's deputy prime minister Prajin Juntong, December 3, 2017, at the opening of the World Internet Conference in Wuzhen.

234 *Freedom on the Net 2017: China Country Profile*, Freedom House, November 2017 (https://freedomhouse.org /report/freedom-net/freedom-net-2017).

235 "China Calls for Tech Collaboration While Criticizing U.S. Action," Bloomberg News, October 20, 2019, 07:17, (MESZ https://www.bloomberg.com/news/articles /2019-10-20/china-calls-for-tech-collaboration-while-criticizing-u-s-action).

236 "Ethiopia: Telecom Surveillance Chills Rights," Human Rights Watch, March 25, 2014 (https://www.hrw.org /news/2014/03/25/ethiopia-telecom-surveillance-chills -rights).

237 Diana Adams, "Real Life Huawei Collaborative Smart City and Safe City Solution in Action," *Medium*, November 15, 2017 (https://medium.com/@adamsconsulting/real -life-huawei-collaborative-smart-city-and-safe-city-solution-in-action-e340657d89f0).

238 "Chinese technology brings falling crime rate to Ecuador," Xinhua, January 19, 2018 (www.xinhuanet.com /english/2018-01/19/c_136908255.htm).

239 *Freedom on the Net 2018. The Rise of Digital Authori-*

tarianism, Freedom House, November 2018 (https://freedomhouse.org/report/freedom-net/freedom-net-2018/rise-digital-authoritarianism).

240 Valentin Weber, "The Worldwide Web of Chinese and Russian Information Controls," S. 21 (https://www.opentech.fund/news/examining-expanding-web-chinese-and-russian-information-controls/).

241 R.S. Foa, A. Klassen, M. Slade, A. Rand, and R. Collins, "The Global Satisfaction with Democracy Report 2020," Cambridge, United Kingdom: Centre for the Future of Democracy, January 2020 https://www.bennettinstitute.cam.ac.uk/publications/global-satisfaction-democracy-report-2020.

242 Yascha Mounk and Roberto Stefan Foa, "This Is How Democracy Dies: A new report shows that people around the world are collectively losing faith in democratic systems," *The Atlantic*, January 29, 2020 (https://www.theatlantic.com/ideas/archive/2020/01/confidence-democracy-lowest-point-record/605686).

243 Dana Heide, Till Hoppe, Stephan Scheuer, and Klaus Stratmann, "EU ambassadors band together against Silk Road," *Handelsblatt Global*, April 17, 2018 (https://global.handelsblatt.com/politics/eu-ambassadors-beijing-china-silk-road-912258).

244 Jiang Jie, "China's Belt and Road Initiative ushers in "Globalization 2.0": experts," *People's Daily Online*, April 12, 2017 (http://en.people.cn/n3/2017/0412/c90000-9202011.html).

245 Martin Hála, "China's gift to Europe is a new version of crony capitalism," *The Guardian,* April 18, 2018 (https:// www.theguardian.com/commentisfree/2018/apr/18/ chinese-europe-czech-republic-crony-capitalism).

246 Kathrin Hille and James Shotter, "Czech university mired in Chinese influence scandal," *Financial Times,* November 11, 2019 (https://www.ft.com/content/ba8645ca-016c -11ea-b7bc-f3fa4e77dd47).

247 "Xifang guojia weihe fangbuxia zhanzheng dabang?" ("Why do the western countries not lay down the cudgel of war?"), Xinhua, November 3, 2016 (http://www.xin huanet.com/world/2016–11/03/c_1119841600.htm).

248 Ai Jun, "Governments worldwide explore internet management," *Global Times,* April 1, 2018 (http://www.glo baltimes.cn/content/1096201.shtml).

About the Author

KAI STRITTMATTER has studied China for more than 30 years. For over a decade he lived and worked in Beijing as a correspondent for the German national broadsheet *Süddeutsche Zeitung*, making the decision to leave the country prior to publication of this book, which became an instant bestseller. He is the author of *China A to Z*.

HARPER LARGE PRINT

We hope you enjoyed reading
our new, comfortable print size and found it
an experience you would like to repeat.

Well – you're in luck!

Harper Large Print offers the finest in
fiction and nonfiction books in this same larger
print size and paperback format. Light and easy to read,
Harper Large Print paperbacks are for the book lovers
who want to see what they are reading without strain.

For a full listing of titles and
new releases to come, please visit our website:
www.hc.com

HARPER LARGE PRINT